H. Lawless   11/85

ACS SYMPOSIUM SERIES **289**

Characterization and Measurement of Flavor Compounds

Donald D. Bills, EDITOR
U.S. Department of Agriculture

Cynthia J. Mussinan, EDITOR
International Flavors and Fragrances

Developed from a symposium sponsored by
the Flavor Subdivision of
the Division of Agricultural and Food Chemistry
at the 188th Meeting
of the American Chemical Society,
Philadelphia, Pennsylvania,
August 26–31, 1984

American Chemical Society, Washington, D.C. 1985

Library of Congress Cataloging in Publication Data

Characterization and measurement of flavor compounds.
(ACS symposium series, ISSN 0097-6156; 289)

"Developed from a symposium sponsored by the Flavor Subdivision of the Division of Agricultural and Food Chemistry at the 188th Meeting of the American Chemical Society, Philadelphia, Pennsylvania, August 26–31, 1984."

Includes bibliographies and indexes.

1. Flavor—Analysis—Congresses.

I. Bills, Donald D., 1932–   . II. Mussinan, Cynthia J., 1946–   . III. American Chemical Society. Division of Agricultural and Food Chemistry. Flavor Subdivision. IV. American Chemical Society. Meeting (188th: 1984: Philadelphia, Pa.) V. Series.

TP372.5.C46   1985   664   85–22913
ISBN 0–8412–0944–8

ACS Symposium Series

M. Joan Comstock, *Series Editor*

Advisory Board

FOREWORD

The ACS SYMPOSIUM SERIES was founded in 1974 to provide a medium for publishing symposia quickly in book form. The format of the Series parallels that of the continuing ADVANCES IN CHEMISTRY SERIES except that, in order to save time, the papers are not typeset but are reproduced as they are submitted by the authors in camera-ready form. Papers are reviewed under the supervision of the Editors with the assistance of the Series Advisory Board and are selected to maintain the integrity of the symposia; however, verbatim reproductions of previously published papers are not accepted. Both reviews and reports of research are acceptable, because symposia may embrace both types of presentation.

CONTENTS

v

PREFACE

BECAUSE THE PERCEPTION OF FLAVOR involves human sensory organs, the characterization and measurement of flavor compounds require more than qualitative and quantitative chemistry. Both chemical and sensory studies are necessary to develop a full understanding of the role of a compound in food flavor. With the introduction of gas chromatography (GC) in the late 1950s, early investigators of flavor combined instrumental and sensory techniques by positioning their noses near the effluent end of the column when separating distillates or extracts of food. Not surprisingly, major components detected by the GC detector often had little or no perceptible odor, although some minor components were potently odorous. Familiarity with the odors of several hundred authentic compounds permitted these early workers to assign tentative identifications to many of the unknown compounds that eluted from their GC columns, and conclusive identifications by rigorous chemical or instrumental methods were often just necessary formalities. The use of the nose as an auxiliary GC detector is still of value, even though the sophistication of instrumental and sensory techniques for flavor compounds has increased greatly.

In the broadest sense, flavor is a perceptual complex, an integration of information received by the brain from the five major human senses. The perception of the flavor of a food or beverage is influenced by appearance, textural or "mouth-feel" properties, sounds associated with mastication in the oral cavity, and sensations derived from the two chemical senses: taste and smell. Although taste and smell usually are most important to flavor and are the senses influenced by flavor compounds, the senses of sight, touch, and hearing must not be discounted. For example, American consumers expect potatoes to be white, but Europeans enjoy yellowish potatoes with a higher content of carotenoids; mashed potatoes should have a uniform, nongrainy texture; and potato chips must produce an audible crunch in the mouth. The purpose of this volume is not to explore the optical and structural characteristics of food, but one must bear in mind the modifying influence that the senses other than olfaction and gustation have on the perception of flavor.

The number of compounds that may contribute to flavor is large. In general, those that influence the sense of smell must be volatile enough to reach the olfactory organ in the upper nasal cavity, whereas those that influence the sense of taste need not be volatile but must have at least a degree of solubility in water. Volatility and solubility do not assure that a compound will be perceived as having an odor or taste. The human sensory apparatus is quite "blind" to many compounds and has variable sensitivity to those that elicit the sensation of flavor. Some compounds are organoleptically detectable and contribute to food flavor at the parts-per-billion level, whereas others must be present at concentrations exceeding a part per thousand to serve as flavor stimuli. The natural flavor of foods typically depends upon the presence of a large number of flavor compounds of several chemical classes, with varying volatilities and polarities, in both large and small concentrations, and in the proper proportions to each other to elicit the characteristic sensation of flavor for a given food. Thus, the characterization and measurement of flavor compounds provide a formidable and continuing challenge.

The authors of this book provide an update of the methodology used in flavor research. Many improvements in instrumental sensitivities and capabilities have accrued in recent years, and state-of-the-art instrumentation and instrumental techniques for flavor analyses compose a substantial portion of this volume. New methods for extracting, derivatizing, and otherwise manipulating flavor compounds are another important part of this book, as are the chapters that deal with sensory evaluation. As editors, we are grateful to the authors for their contributions and to our respective employers for their support of our effort.

DONALD D. BILLS
U.S. Department of Agriculture
Philadelphia, PA 19188

CYNTHIA J. MUSSINAN
International Flavors and Fragrances
Union Beach, NJ 19118

June 24, 1985

Sensory Evaluation of Food Flavors

M. R. McDaniel

Department of Food Science and Technology, Oregon State University, Corvallis, OR 97331-6602

Many changes have occurred in the sensory analysis of flavor in the past half century, beginning with the phasing out of the inappropriate term, "Organoleptic," and the utilization of more appropriate terms to describe the field of study such as "Sensory Analysis" or "Sensory Science." In the food and flavor industries, bench scientists using unsophisticated methods have been replaced by highly educated specialists (M.S. and Ph.D.'s in Food Science and Psychophysics) directing large sensory groups. These specialists utilize highly trained judges and sophisticated test designs and analysis to solve sensory problems. Consumer testing in most industry applications now is done with consumers. Even with the continuing development of instrumentation to replace the human judge, sensory analysis continues to expand its contribution to flavor analysis.

Sensory evaluation (sensory science) is a scientific discipline that concerns the presentation of a stimulus (in this case a flavor compound, a flavor, or flavored product) to a subject and then evaluation of the subject's response. The response is expressed as, or translated into, a numerical form so that the data can be statistically analyzed. The sensory scientist then collaborates with the research or product development team to interpret the results and to reach decisions. Sensory scientists stress that decisions, such as product formulation, are made by people, not by the results of a sensory test, although such results may provide powerful guidance in the decision-making process.

Sensory science is unique in that it requires human subjects. This in itself creates challenges, some of which will be discussed in this paper. The sensory scientist, often working as a part of a research team, also is unique because training in a number of fields is necessary to the success of the program. The training of sensory scientists has not proceeded as rapidly as has the appreciation of and need for sensory scientists in the flavor and

0097-6156/85/0289-0001$06.00/0

food industries. In the past, virtually anyone, regardless of
training and background, might be handed a sensory methods manual,
and told that they were in charge of the sensory program. For-
tunately this rarely takes place today. Sensory scientists are
still in relatively short supply, but there are dozens of highly-
trained men and women who are currently directing sensory programs
in both flavor and fragrance companies and their clients' companies.
A sensory scientist working in the area of flavor would be expected
to have a background in food science and at least basic knowledge
in the areas of physiology, psychology and statistics.

History

Forss (1) reviewed the relationship between sensory analysis and
flavor chemistry, and provided a discussion of sensory characterization of flavor. Williams et al. (2) also addressed the problem
of relating the many known flavor compounds to what is actually
perceived by the individual on a physico-chemical basis. This
chapter will center more on reviewing advances in sensory methodology as they have been adapted by the flavor and food industry.
    Moskowitz (3) in his book "Product Testing and Sensory
Evaluation of Foods" reviewed the history of sensory evaluation
beginning with the study of psychophysics. Psychophysics is the
study of the relationship between a physical stimulus and a subject's psychological response to that stimulus. Very early work
was done in Germany by E.M. Weber and G.T. Feckner over a hundred
years ago. They were seeking quantitative laws of human perception,
specifically how to measure our ability to discriminate. The study
of psychophysics advanced rapidly over the following years and
along with it, so did testing methodology. As the psychologists
developed and tested their methods, food scientists and sensory
analysts borrowed those methods and applied them to the study of
food. Moskowitz (3) further observed that sensory analysis
evolved greatly during World War II. Much of this work was done
by the U.S. Army Quarter-Master Corps, led by Peryam and Pilgrim,
developers of the 9-point Hedonic scale.
    Also during this time, Stevens (4), a psychologist at Harvard
University, was working on the psychophysics of hearing, and his
work led to the development of a theory of scales of measurement,
and ultimately to the development of the power law. The power law
is generally accepted to define the relationship between psychological response and physical stimulus. The log of this equation
is readily recognizable as the equation for a straight line,
therefore plots of psychological response versus physical stimulus
concentration on a log-log plot result in a straight line. These
plots, depending upon which sensation is being measured, be it
electrical shock, brightness or sweetness, have different slopes,
and the slope is defined as the power of function. Food scientists
and psychophysicists have applied these laws to food evaluations
and have obtained valuable information.
    Although knowledge of the power law has been available for
many years, food scientists have used this information only recently to define how the various sensory qualities in foods

change, with changes in concentration. It is surprising that many
food companies with only a few major products still do not know
what the sensory properties of their products are or how they
change with changes in formulation.

Scaling Methodology

A new method called magnitude estimation, a form of ratio scaling,
also emerged from Stevens (4) work. To estimate the magnitude of
some stimulus, panelists simply assign a number to reflect the
stimulus intensity. Prior to the use of magnitude estimation, food
scientists and other psychologists used rather arbitrary category
scales in order to quantify perceptions. The scales could vary in
length, from a three point scale to perhaps a nine point scale or
even larger, usually with each scale point representing some
intensity and designated as small, moderate, large, extreme or by
some other descriptive word. However, because of the nature of
the scale itself and the way in which it was used by the judge, if
one plotted the relationship between psychological response and
stimulus intensity, one usually would not obtain a linear response.
A log-log plot of stimulus intensity versus psychological response
for magnitude estimation usually results in a straight line.
     Many studies have attempted to compare the results of magni-
tude estimation versus some of the more standard scaling techniques
or semi-structure or unstructured line techniques and different
results have arisen from these various studies (5-8). Based on my
experience, the more important consideration is the degree of
training of the panel utilized in trained panel work rather than
the type of scale that is used. However, magnitude estimation
results do tend to show more differences when the differences are
very small than do other methods.

Consumer Testing

Perhaps the biggest advance in consumer testing is the use of
consumers instead of trying to obtain consumer data from people
who do not use the product or who are too close to the product
(a company's own workers). It once was standard procedure to do
in-house consumer testing with company employees evaluating the
products that produced their paychecks. This was not an unbiased
sample and led to many expensive mistakes by industry. Most
companies now use a marketing organization to do central location
testing or home placement testing of their products. This is not
to say that all in-house consumer testing is incorrect. The
concern is that definite risks are involved, and one must understand that misinformation may result. Some very valuable
directional information can be gained in the right circumstances
by using in-house panels. Validity of this information can only
be judged through experience with the individual industries and
products involved.
     One of the most important things to learn in sensory evalua-
tion is that "experience" is critical in making methodology
decisions. Years of sensory testing on one product or one product

line yields valuable information that should be used by the
researcher and the product developer in making decisions regarding
their current line. However, using the same methodology just
because it's been used for years and years is not necessarily a
wise decision. Some methodologies are incorrect or insensitive,
or the words used in scales have little relative meaning. One
should not be afraid to test new methodology versus old methodology
to see if valuable information can be gained by switching to
another method.

The food industry also has learned that they must test
consumers in large numbers. However, unfortunately, there is no
magic "large" number, and the number of judgments that is
decided upon must be based, again, on the risk they are willing
to take. Another valuable lesson learned over the years in
testing products is that consumers are all very different, and
these very different segments of the population, where a product
market exists, must be tested. If a company is considering
changing the formulation of a product, they must test current
consumers of the product to see if the change makes a difference
to them. They also may wish to test consumers of their product
versus consumers of a competitor's product, but this yields
different information. If the consumers of their products are
children, they must test children, and perhaps also the parent
who purchases the product. If the consumers of a product are on
special diets, people on special diets should be the subjects of
the test. For example, a non-gluten bread may be highly appreci-
ated by one on a gluten-free diet, but totally unacceptable to
those who have no need to restrict gluten in their diets.

Experience has shown that the order of sample presentation,
when a number of samples are to be presented at one time, is very
important. The first product may strongly bias the evaluation
of the product following it, and one may find a significant order
effect with some products. The sample presentation order in any
test must be balanced or randomized. In analyzing data from such
tests, one should consider the average score of one sample
presented first versus its average score when presented second.
The fact that your competitor's product presented first somehow
causes judges to score your product presented second lower, can
be valuable information. What was it about the competitor's
product or about your product that caused this difference? The
amount of sample is also critical. Very often in a taste test,
people are given just a "taste" and this may not be enough for
them to truly evaluate the product. For example, when testing
soup, where the flavor will build up, mouthful after mouthful,
it may be necessary to have each panelist consume an entire bowl
of soup. The first impression gained from one or two bites of
a product may be totally different than the final impression after
consuming an entire serving. This can be true with many products,
especially highly flavored products.

Difference Testing

Difference testing has not changed greatly over the years. The
triangle test and duo-trio test remain popular and well-accepted,
although much effort has been extended to prove one better than
the other. The best advice is to use the one that fits your
test situation.

When conducting one of the pure difference tests, the
triangle or duo-trio test, one must realize that there are two
distinct types, one being a similarity taste and the other being
a difference test. In the case of the similarity test the
experimental samples are actually different from the control.
However, the purpose is not to create a perceptible difference
but to effect a cost reduction or to change suppliers of raw
materials without changing the product identity. The actual
goal of such a project is to change the product without influencing consumer perception and acceptance of it. The statistical
error of concern here is the beta error which is the probability
of concluding that the two samples are not different when they
are different. In such testing, it is desirable to keep the beta
risk very low, but keeping both the alpha and beta risks low is
difficult. When alpha is low, beta tends to be high, and when
beta is low, alpha tends to be high. The only way to insure that
both errors are small is to test a very large number of people.

In a difference test, rather than a similarity test, one
would intentionally make the samples different and then ascertain
whether judges could detect the difference.

Product Drift

A major concern in the food industry is product drift or subtle,
step-wise changes that take place in the product over time.
Product drift can occur when the original (first) product is
tested and found not different from the new (second) product.
The second product is changed in yet another way, but is found
not to be different from the newest (third) product. However,
the third product may be different from the first product. The
best way to avoid product drift is to intimately know a product.
This means knowing exactly what sensory characteristics are
present in your product and at what intensities. This type of
sensory analysis requires the use of descriptive analysis
techniques. Because descriptive analysis involves using a
trained panel, developing a set of descriptors, and rating
their intensities, it can be very expensive, but so can product
drift.

Descriptive Analysis

One of the most exciting developments in sensory evaluation over
the past decades has been the emergence and popularity of
descriptive analysis or the use of highly trained panels to
describe the sensory characteristics of foods. This is perhaps
the most important development in sensory evaluation methodology.

Only when products are described in detail and the intensity of
descriptors rated can true product differences or drift be
noted.
      The type of descriptive analysis chosen should be based on
the variety of products produced by the company. If a company
produces only one type of product, a more limited or specific
training such as quantitative descriptive analysis described by
Stone et al. (9) may suffice. If a company has a large variety
of products, flavor profile training established by the Arthur D.
Little Company may be more efficient. Many laboratories combine
the best of both methods and develop their own method. The key
to success with either method is panel training and the establish-
ment of appropriate terminology.
      To illustrate descriptive analysis, I will draw from both
the wine and beer industry. Oregon State University's Sensory
Science Laboratory, located in the Department of Food Science and
Technology, is heavily involved in wine and beer research. The
principle problems and solutions in the sensory analysis of wine
and beer should be transferable to other products. Common wine
descriptors, such as soft, hard, fat, are ambiguous. What do
soft or hard mean when referring to wine? The goal of descriptive
analysis is to use precise terms, even referring to specific
chemical entities when possible. In the wine industry, objective
sensory analysis must overcome the historical romance of wine.
A classic example is the following description of a wine, "It's
a naive domestic burgundy without any breeding but I think you'll
be amused by its presumption." Such a description obviously
lacks meaningful sensory terms that convey an impression of the
wine's aroma and taste. In work on Pinot Noir qualities in our
laboratory, a set of sensory descriptors were developed to aid in
evaluating the effect of several processing variables, Henderson
and McDaniel (10). A trained panel used the ballot shown in
Figure 1 to describe wine produced by different malolactic
cultures. There are several ways to display this type of information. The QDA method joins descriptor intensity points together
to visually display difference. This works very nicely for two
to three comparisons. When one has more than two to three samples
to compare, other types of statistical analyses and methods of
displaying the results may be employed. This will be covered in
a later section.

Statistical Analysis

Sensory scientists rely greatly on statistical analysis to aid
in interpretation of data. They also continue to argue endlessly
about what is correct and incorrect. Some of the questions that
are posed include: which is the right analysis; are the assumptions
of the test being violated; is the data good enough in the first
place to have statistical analyses applied to it.
      Some of the most interesting advances in sensory analysis in
the last 20 years have been in the area of statistical evaluation
of the results. Multivariate analysis is an example of a new
type of statistical analysis applied to food system. Multivariate

NAME _____

DATE _____ SAMPLE # _____

Using the 9-point intensity scale shown below, rate each sample for all attributes listed.

FOR AROMA ONLY

1 - none
2 - threshold
3 - slight
4 - slight to moderate
5 - moderate   Overall Intensity _____
6 - moderate to large
7 - large
8 - large to extreme
9 - extreme

1st tier	2nd tier	3rd tier	
Fruity _____	Citrus _____	grapefruit	_____
	Berry _____	blackberry	_____
		strawberry	_____
		raspberry	_____
	Tree Fruit _____	cherry	_____
	Dried Fruit _____	strawberry jam	_____
		raisin	_____
		fig	_____
		prune	_____
Spicy _____	Spicy _____	black pepper	_____
		cloves	_____
Vegetative _____	Canned/cooked _____		
Earthy _____			
Caramelized _____	Caramelized _____	honey	_____
		buttery	_____
		butterscotch	_____
Chemical _____	Pungent _____	Ethanol	_____
	Sulfur _____		
Microbiological _____	lactic _____		

Figure 1. Ballot used by trained panel for evaluation of wine.

analysis is used when observations are collected on many different variables, and is particularly helpful when one may be overwhelmed by the sheer bulk of the data that has been collected. There are several types of multivariate analysis, 1) principal component analysis; 2) factor analysis; 3) cluster analysis; 4) discriminate analysis. Principle component analysis and factor analysis are very similar, a major difference being that factor analysis requires assumption of normality. Both methods take a list of variables and reduce this to a smaller number of factors (made up of these original variables). Such analyses have been applied in wine research (13, 14). An example would be in the evaluation of wine where sensory and/or analytical measurements have been taken on many samples. The goal would be to reduce the number of variables necessary to actually show meaningful differences between the wine samples. Factor analysis with vector loading is helpful in this situation. The loadings tell you how each variable fits under each factor.

Cluster analysis allows one to see how close samples are together in a multidimensional space. An analogy can be drawn to playing cards, which may cluster in obvious ways by suit or card, or by other means inherent in some game playing rules. Cluster analysis is generally a stepwise progression. An example, from hop variety research involving eight varieties and many samples of each, is an appropriate subject for application of this technique (15). The samples clustered near one side would be most similar. Also, any two compounds whose concentration ratios were relatively constant in all or most samples would have a high similarity value and would cluster on the same side.

With discriminate analysis one is concerned about how observations differ and one sets the rules to distinguish between populations. The result is a type of classification or sorting of observation into groups. For example, an unknown wine variety may be classified among known varieties.

Another example is the application of response surface methodology (16, 17). The response is some function of the design variables in the test and all of the variables are well controlled and precisely measurable. In order to visualize response surface methodology, imagine that you are viewing a mountain on the horizon. This would be a single variable x and its response, y showing the maximum. Imagine adding a second dimension, 2, coming straight out at you, to give the mountain dimension, and then starting at the very top of the mountain, taking slices of that mountain on a horizontal axis at equal response lines. If you then look down on that slice of the mountain, you will see circles, the smallest inner circle equaling the largest response. Response surface procedures are not used to understand the mechanism of the underlying system, but rather to determine what optimum operating conditions are or to determine a region of the total space of the factors in which certain operation specifications are met.

Conclusion

Almost everyone is now utilizing the computer for statistical
analysis of sensory data. Some laboratories also are using
computers to gather the data as well (18). A computerized
sensory system would benefit most laboratories by freeing
workers from laborious data entry and analysis. Also, it
would allow for a more thorough analysis of the data. It
should not replace inspection of the raw data by the sensory
scientist, but allow this to occur more easily.

The field of sensory evaluation has matured over the years.
We have learned through expensive mistakes to rigorously control
test situations to obtain valid data and to analyze the data as
thoroughly as possible to maximize understanding of products.
I believe the future of sensory evaluation will involve an
expansion of the use of descriptive analysis in many different
situations, such as in plant quality control, as well as product
development and research applications. Because of the increased
competition in the flavor industry, flavor companies are
increasingly expanding their sensory work and sensory capabilities. This is necessary, not only for the flavor company
to understand the products they are producing but to be able
to satisfactorally service their client companies.

Literature Cited

1. Forss, D.A., in Flavor Research Recent Advances. (S.R.
    Tannenbaum and P. Watson, ed.). Marcel Dekker, Inc.
    1981. p. 125
2. Williams, A.A.: Lea, A.G.H.; Timberlake, C.F., in Flavor
    Quality Objective Methods. (R.A. Scanlan, ed.) ACS
    Symposium Series No. 51, Washington, D.C., American
    Chemical Society, 1977. p. 71.
3. Moskowitz, H.R. Product Testing and Sensory Evaluation
    of Foods. Food and Nutrition Press, Inc. Westport, CT.
    1983.
4. Stevens, S.S. Science 1946, 103, 677-678.
5. Moskowitz, H.R. and Sidel, J.L. J. Food Sci. 1971, 36,
    677-680.
6. McDaniel, M.R. and Sawyer, F.M. J. Food Sci. 1981, 46,
    178-181.
7. Giovanni, M.E. and Pangborn, R.M. J. Food Sci. 1983, 48,
    1175-1182.
8. Shand, P.J.; Hawrysh, A.J.; Hardin, R.T.; and Jeremiah, L.E.
    J. Food Sci. 1985, 50, 495-500.
9. Stone, H.; Sidel, J.; Oliver, S.; Woolsey, H.; and
    Singleton, R.C. Food Technology 1974, 28(11), 24-34.
10. Henderson, L.A. and McDaniel, M.R. Personal communication,
    1985.
11. Ennis, D.M.; Boelens, H.; Haring, H.; and Bowman, P.
    Food Technol. 1982, 36(11), 83-90.

12. Johnson, R.A. and Wichern, D.W. Applied Multivariate Statistical Analysis. Prentice-Hall, Inc. Englewood Cliffs, New Jersey, 1982.
13. Wu, L.S.; Bargmann, R.E.; and Powers, J.J. J. Food Sci. 1977, 42, 944-952.
14. Williams, A.A. J. Inst. Brew. 1982, 88, 43.
15. Stenroos, L.E. and Siebert, K.J. ASBC Journal 1984, 34, 55.
16. Henika, R.G. Food Technology 1982, 36(11), 96-101.
17. Myers, R.H. Response Surface Methodology. Allyn and Bacon, Inc. Boston, 1984.
18. Brady, P.L.; Ketelsen, S.M.; and L.J.P. Ketelsen. Food Technology 1985, 39(5), 82-88.

RECEIVED July 29, 1985

2

Substances That Modify the Perception of Sweetness

Michael A. Adams

Monell Chemical Senses Center, University of Pennsylvania, Philadelphia, PA 19104

Human taste response is modified by several plant-derived
substances. The detergent sodium dodecyl sulfate, as
well as triterpene saponins from the leaves of several
plant species (most notably Gymnema sylvestre and
Ziziphus jujuba) will temporarily inhibit the sweet taste
sensation in man; the duration of the effect being about
one hour for G. sylvestre and about fifteen minutes for
Z. jujuba. The mechanism of action seems to be related,
in part, to the surfactant properties of the materials.
Structures of the modifiers and possible mechanisms of
action are discussed.

The number of substances known to modify sweet taste is quite
small. Miraculin, a substance from the fruit of Synsepalum
dulcificum, has the ability to make sour stimuli taste sweet. The
fruits of this West African shrub ("miracle fruit") have been used
by natives to improve the flavor of maize bread, sour palm wine,
and beer (1). The taste modifying effect is of quite long
duration, sometimes lasting for more than three hours. Miraculin
is a glycoprotein with a molecular weight of about 44,000 (2). Not
much is known about its structure or its mode of action, but it has
been proposed (3) that this substance binds to the taste cell
surface, where its effect is then manifested. It was further
suggested that exposure of the taste receptors to acid contained in
a food or beverage causes a conformational change in the membrane
enfolding the receptor, which allows the sugar moieties bound to
miraculin to stimulate the sweetness receptor. The speed of
response of miraculin is on the order of milliseconds, and this is
taken as evidence that the sweetness receptor lies on the outside
of a taste cell, since the speed of response is apparently too fast
to allow for transport into the cell interior.
    Additional evidence that the exterior surface of the taste
receptor cell plasma membrane is the location of the sweet receptor
is provided by the action of the chemostimulatory proteins,
monellin and thaumatin. Monellin occurs in the fruit of the
African serendipity berry (Dioscoreophyllum cumminsii), and
thaumatin is found in the fruit of Thaumatococcus daniellii, also

0097–6156/85/0289–0011$06.00/0

from Africa. Both of these proteins have an intensely sweet
taste. Studies have shown that they are essentially carbohydrate-
free, in contrast with miraculin. Monellin is a single peptide
chain with a molecular weight of about 11,000, and thaumatin,
believed also to be a single peptide chain, has a weight of
approximately 21,000. Evidence has been presented indicating that
the tertiary structures of these proteins are critical to the sweet
taste (4), and Cagan (5) has suggested that the size of these
molecules renders it unlikely that they are being transported
inside the taste receptor cell. It is reasonable to conclude that
their taste-stimulating effects are most likely being manifested at
the cell surface.

The effects of miraculin, monellin and thaumatin, taken
together, provide evidence that the sweetness receptor site is
located at the surface of the taste receptor cell, in or in close
proximity to, the plasma membrane.

Sweetness Inhibitors

There are only a few sweetness inhibitors known. This article will
focus on three: sodium dodecyl sulfate (SDS), the gymnemic acids
(GA) and the ziziphins (ZJ). Each of these substances has the
ability to diminish or eliminate the ability to recognize
sweetness. The intensity and duration of the effect varies with
the inhibitor.

The best-known sweetness inhibitor is sodium dodecyl sulfate
(SDS), also known as sodium lauryl sulfate. This substance is a
twelve carbon surfactant that is quite commonly used as a detergent
in toothpaste. The observation is often made that after brushing
one's teeth, the taste of orange juice is unusually bitter. This
has been ascribed to the presence of SDS in the dentifrice (6).
The duration of the effect, as measured in psychophysical
experiments, is very brief - on the order of minutes, and the
suppression effect is not complete.

A substance isolated from the Indian shrub Gymnema sylvestre,
has a profound ability to reduce perceived sweetness of sugar
solutions. The effect was noticed over a century ago when two
British inhabitants of an Indian village found that, after chewing
the leaves of G. sylvestre, the sweetness of their tea disappeared
(7). The sweetness suppressing activity is due to a mixture of
several triterpene saponins which have collectively been termed the
gymnemic acids. For most people exposed to the effects of GA,
sweetness suppression is complete and the effect lasts for about an
hour.

Much work was done in the 1960's by Stocklin, Sinsheimer and
their coworkers to isolate, purify, and elucidate the structures of
these taste inhibitors (8-11). Many substances without antisweet
activity were isolated from the leaves of G. sylvestre, including
hydrocarbons, stigmasterol, β-amyrin, lupeol and phytol, among
others. A crude antisweet extract was, however, produced by
mineral acid precipitation of an aqueous leaf extract. This
extract, when dried, accounts for about 10% of the leaf material.
Chloroform extraction was then used to remove some remaining
inactive material. Pfaffmann (12) determined that the mixture of
taste modifying substances was glycosidic in nature, as glucose,

arabinose, and glucuronolactone were obtained upon hydrolysis. The
hydrolysate was found to have no antisweet effect. Yackzan (7)
postulated that the sweetness-inhibiting gymnemic acid must be a
saponin, based on its foaming in solution, ability to produce red
blood cell lysis, and its glycosidic nature.

Extraction of the crude leaf fraction with acetone removes all
physiologically active material. This acetone-soluble material was
chromatographed on silica gel by Stocklin (9) to give the various
closely-related gymnemic acids. Further purification by Sinsheimer
and Subba Rao led to the structural identification of gymnemagenin
(I) and gymnestrogenin (II) as the aglycones of the gymnemic acids
(Figure 1). The major taste active substance appears now to be
gymnemagenin (I), esterified with glucuronic acid. This saponin is
called gymnemic acid A, (III, Figure 2). The identity of the R
groups is presently unknown, although claims for formic, acetic,
isovaleric and tiglic acids have been made (10, 13, 14). No
further structural work has been done with the gymnemic acids for
the last several years.

Once a purified gymnemic acid became available, much psychophysical work was done to understand the nature of the sweetness
inhibition effect. The work of Bartoshuk and co-workers
illustrates the course taken (15). The results of a typical
experiment are shown in Figure 3. The sweetness of a sucrose
solution was almost completely suppressed after holding a gymnemic
acid solution in the mouth for a few seconds. Further experiments
were carried out to determine the effect of gymnemic acid on the
other taste qualities (sour, bitter and salty). No effect of
gymnemic acid on these tastes was observed. Early work with GA
extracts had produced an apparent inhibition of bitter taste, but
this effect was later attributed to cross-adaptation to the taste
of the crude leaf extract, which was itself quite bitter.
Experiments with refined (and tasteless) extracts showed no
bitterness suppression.

A third group of sweetness inhibiting substances has recently
been isolated from the leaves of the Middle Eastern tree, Ziziphus
jujuba. This material, consisting of a mixture of 15-20 triterpene
saponins, has activity similar to that of the gymnemic acids, but
the duration of the effect is much shorter, on the order of 15
minutes for the average subject. The duration of suppression
varies from subject to subject, and is usually not complete. The
isolation, purification, and structural characterization of these
sweetness inhibitors, which have been termed the ziziphins, is
under investigation in our laboratory and some of our recent
results are described below. Psychophysical and preliminary
chemical studies were carried out with Z. jujuba (ZJ) extracts by
Meiselman and co-workers (16). Their chromatographic
investigations showed that the gymnemic acids were not present in
ZJ leaf extracts, but their observations of the physical properties
of leaf extract solutions did indicate that saponin-like substances
were present. The effect of ziziphin treatment of the tongue on
sweet, sour, bitter, and salty tastes were examined. As with the
gymnemic acids, ZJ only reduced the intensity of sweet taste.

A more detailed investigation of the ziziphins was carried out
by Kennedy and Halpern (17). Included in their work was a
phylogenetic analysis of Z. jujuba and Gymnema sylvestre (Figure

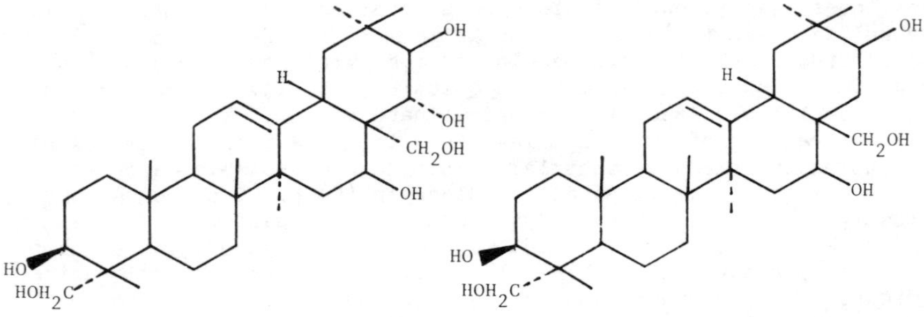

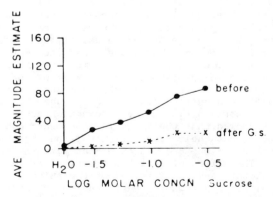

Figure 1.   Structure of gymnemagenin (I) and gymnestrogenin (II).

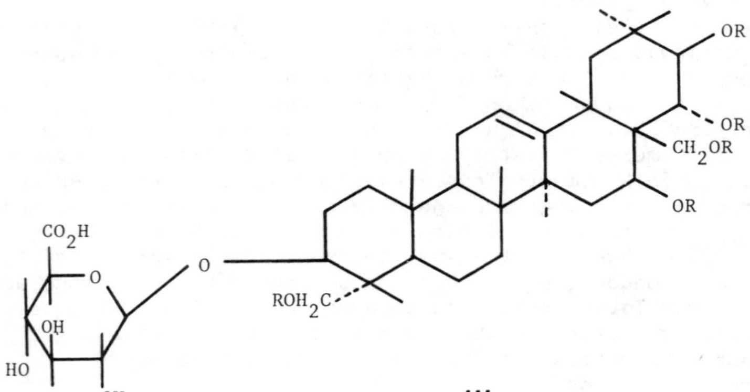

Figure 2.   Structure of gymnemic acid A, (III), the most abundant gymnemic acid.

Figure 3.   The effects of _Gymnema sylvestre_ on the taste intensity of sucrose. Reproduced with permission from Ref. 15. Copyright 1969, Rockefeller University Press.

4). From this analysis it is clear that the two species are not
closely related, having diverged at the subclass level. This
provides botanical support for the observation of Meiselman (16)
that gymnemic acids are not detectable by thin layer chromatography
in (ZJ) leaf extract. It is now clear that these two plants,
although widely separated phylogenetically, manufacture substances
of different physical constitution, but with similar human
psychophysical properties. Kennedy and Halpern (17) report a
rather difficult procedure for isolating, a purified ZJ fraction.
The separation involved preparing an ethanol-water extract, back
extracting with hexane and diethyl ether to remove impurities, and
then extraction with a chloroform-ethanol solution to give crude
antisweet material. This material was then partitioned between
ether and water and the solids that collected at the ether-water
interface were removed with a Pasteur pipette. After solids
removal, a fresh portion of ether was added and the partitioning
and solids-removal steps were repeated until no more material
collected at the interface (about 15 times). The formation of
emulsions with each extraction and partitioning step contributed
greatly to the time and difficulty necessary to carry out a
preparative-scale isolation.

Our approach to isolation of the sweetness inhibiting
ziziphins is somewhat different (Figure 5). Leaves were collected
in the Fall from two Z. jujuba trees at Longwood Gardens,
Pennsylvania. They were air-dried and then ground to a powder.
Extraction with methanol-water (2:1, repeated three times), gave a
crude plant extract containing the taste-modifying substances.
This material was extracted successively with hexane, ether and n-
butyl alcohol. The sweetness-inhibiting material was contained in
the organic portion of the butanol extract, as determined by human
taste bioassay. Evaporation of the butanol at reduced pressure
produced a tan solid which was then dissolved in the minimum amount
of methanol necessary to effect solution. The methanol solution
was poured into a large volume of diethyl ether and the resulting
precipitate was collected by filtration. This precipitate, the
crude antisweet fraction, represented about 4% of the dry leaf
weight and contained the antisweet substances. Examination of this
fraction by thin-layer chromatography (TLC) revealed the presence
of 15-20 compounds, mostly triterpene saponins with a few flavone
glycosides. A detailed description of our isolation and
purification techniques is beyond the scope of this article, but
the techniques we found most useful are outlined in Figure 6.
Droplet counter-current chromatography (DCC or DCCC) is useful for
separation of the more polar saponins, but we found flash
chromatography (18) and high performance liquid chromatography
(HPLC) to be the most useful for the bulk of our separations.
Preparative-scale TLC has been used for final sample cleanup prior
to spectroscopic analysis.

Acid-catalyzed hydrolysis of any of the saponins contained in
the antisweet fraction resulted in a mixture of sugars, but only a
single aglycone. Upon examination by NMR spectroscopy, this
aglycone was identified as ebelin lactone, IV (Figure 7). Ebelin
lactone was previously isolated from the seed saponins of Z. jujuba
by Shibata and co-workers (19, 20) who showed it to be an artifact
derived from the actual aglycone, jujubogenin (V), (Figure 7) by an

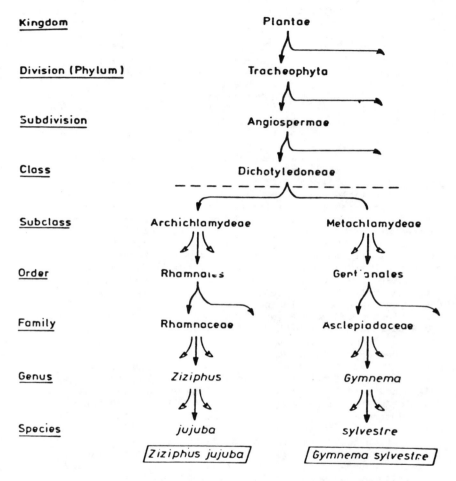

Figure 4.   Phylogenetic classification of Ziziphus jujuba and
Gymnema sylvestre. Lineage. Since separation occurs early, at
the subclass level, these two species do not appear closely
related. Solid arrows indicate single pathways. Open arrows
indicate multiple pathways. Reproduced with permission from
Ref. 17. Copyright 1980, ILR Press, Ltd。

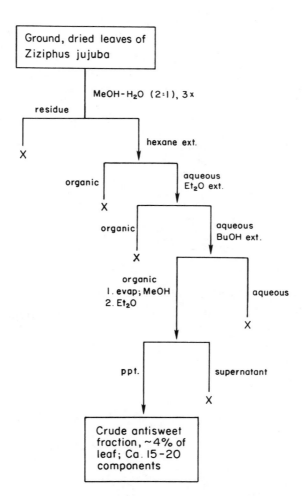

Figure 5.   Scheme for isolation of crude antisweet fraction from the leaves of Ziziphus jujuba.   Taste bioassays were performed at each step to ascertain antisweet activity.   An X indicates that no antisweet activity was detected.

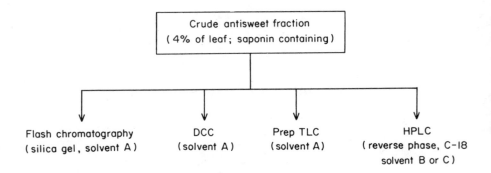

Solvent A = CHCl$_3$: MeOH : PrOH : H$_2$O

Solvent B = MeOH : H$_2$O

Solvent C = CH$_3$CN : H$_2$O

Figure 6.   Techniques used for isolation and purification of antisweet substances from the crude antisweet leaf fraction of *Ziziphus jujuba*.

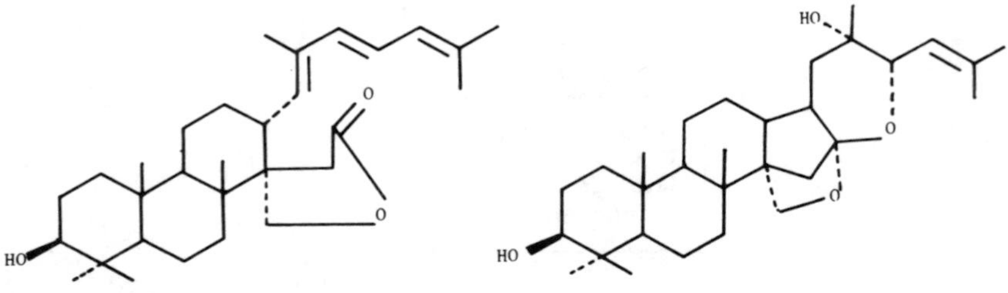

IV   V

Figure 7.   Structure of ebelin lactone (IV) and jujubogenin (V).

acid-catalyzed rearrangement. Thus, it presently appears that all of our saponins are jujubogenin-based, differing only in the type, number, and linkage of sugar moieties. From our hydrolysis studies we have identified glucose, xylose, fucose, arabinose and rhamnose as the sugars of the saponins.

The smallest saponin we have isolated thus far is jujubogenin-glucose-rhamnose and our general approach to structure identification in this system can be illustrated using this molecule as an example. This substance was isolated from the crude antisweet fraction by silica gel flash chromatography. It was purified by HPLC on a reverse-phase silica gel column, followed by prep-scale silica gel TLC. Acid hydrolysis gave an aglycone (shown to be ebelin lactone by NMR and mass spectrometry) and the sugars glucose and rhamnose. The identities of the sugars were verified by thin layer chromatography, gas chromatography, and by comparison with sugar standards.

We have used desorption-chemical ionization mass spectrometry (DCI-MS) (21) to obtain the order of sugar attachment to jujubogenin for this molecule. DCI-MS is an "in-beam" technique in which vaporization-ionization of relatively non-volatile molecules is facilitated. The sample was deposited, as an aqueous solution, on a deactivated rhenium wire filament (Figure 8). After evaporation of the solvent, the filament was inserted into the mass spectrometer's electron beam. The wire was then rapidly heated to effect kinetic desorption of the substance. An ionizing gas (ammonia in this case) was used to facilitate ion production. The sample desorbed very quickly and cleanly from the wire as shown in the reconstructed ion chromatogram in Figure 9 (inset). The mass spectrum is shown in Figure 9. A molecular ion was not seen under the run conditions. An ion corresponding to the ammonia adduct of jujubogenin-glucose is seen at m/z 652, which indicates that rhamnose is the ultimate sugar; glucose is bonded directly to jujubogenin. The ion for the ammonia adduct of jujubogenin appears at m/z 490. This simple experiment permitted assignment of the carbohydrate linkage order, however, it does not say anything about the nature of carbohydrate linkages themselves. This information may be obtained by methylation analysis or circular dichroism techniques, and these determinations are currently being pursued in our laboratory.

Mechanism of Action

Sodium dodecyl sulfate, the gymnemic acids and the ziziphins have all been termed "surface active" taste modifiers because they all possess detergent-like properties. These molecules all have a polar and a non-polar end and they are capable of penetrating the phospholipid membranes that are believed to be components of sweetness receptors. Any speculation about the mechanism of action of these substances must take into account the experimental observations concerning miraculin, monellin, and thaumatin, which were presented at the beginning of this article. Those observations suggested that transport of the modifier to the cell's interior was not occurring and the inhibition effect is manifested at the surface of the cell.

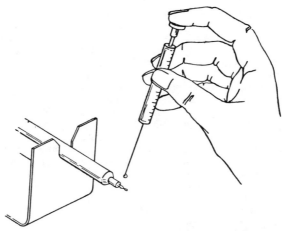

Figure 8。 Preparing a sample for DCI mass spectrometry。
Reproduced with permission from the Finnigan MAT Corporation.
Copyright 1982, Finnigan MAT Corp.

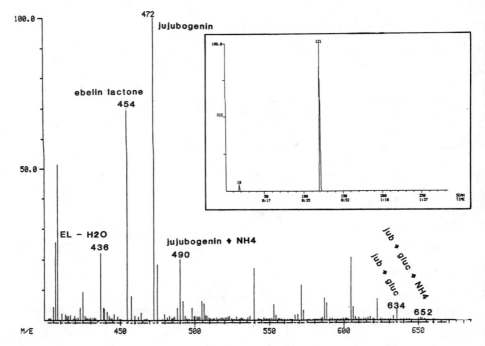

Figure 9. DCI mass spectrum of jujubogenin-glucose-rhamnose.
Conditions: ammonia ionizing gas at 0.25 torr; filament heating
rate 50 deg./sec; 0。3 sec/scan. Inset figure: reconstructed ion
chromatogram of jujubogenin-glucose-rhamnose. Reproduced with
permission from Ref. 29. Copyright 1981, Academic Press, Inc.

The size of the molecules and the speed with which the modifying effect appears argue against action taking place in the interior of a taste cell. DeSimone's work on the physicochemical effects of tastants on phospholipid membranes provides insight into possible mechanisms of modifier action. He has shown that GA and SDS are about equivalent in surfactant strength and that dilute solutions of both substances are capable of penetrating phospholipid monolayers (6). Surfactants are known to affect cell function by disrupting the structure or spatial arrangement of membrane lipoproteins (22). Surface-active taste modifiers are postulated to exert their effects by altering a membrane-related process that relates in some way to the transduction of a taste signal. This disruption of cell structural proteins, or related phospholipid membranes, might be attributed to alterations in the surface pressure (and surface free energy) of a membrane through insertion of the taste modifier into the membrane (23).

Additional insight into the mechanism of action of these inhibitors is provided by the kinetic analysis of Ray and Birch (24). Two facts are known with certainty concerning molecules which produce taste effects: they must be soluble in saliva and they must be able to occupy a receptor site. To disrupt or inhibit taste, one or the other of these characteristics can be interfered with. Of the two, the latter is the one most amenable to influence. From a kinetic standpoint a stimulus molecule combines with a receptor to form a stimulus-receptor complex. This reaction occurs at a certain rate. This complex may then break apart, giving back the receptor and the stimulus molecule; this occurs at another rate. The formation of the stimulus-receptor complex eventually results in the firing of a nerve and the production of a taste sensation. The rates of formation and breakdown of the complex are influenced by temperature. Birch has found that by increasing the temperature of tastant solutions, the perceived intensities of the solutions increase and the response-time profile changes as well (Figure 10). For a given temperature and sucrose concentration, taste sensation plateaus at a certain time, suggesting that the sweetness receptors have become saturated (i.e. that the rate of stimulus-receptor complex formation and breakdown are equal). A line can be drawn on the graph from the beginning point of stimulus perception to the plateau point and the slope of a line so drawn is called the magnitude estimation rate (MER). A plot of the reciprocals of MER values versus corresponding reciprocal sucrose concentrations can be made; one obtains a Lineweaver-Burk type of plot (Figure 11). This plot indicates a low affinity of sugar for receptor and affinity values obtained by this method agree generally with those of Cagan (25), which were determined by a radioactively-labelled sugar binding assay.

In similar fashion, a Lineweaver-Burk type plot for a gymnemic acid inhibited tongue responding to sucrose solutions can be constructed. The appearance of this plot is quite different from that of Figure 12 and is not characteristic for either competitive or non-competitive inhibition. The results of the analysis seem to indicate that inhibition of sweetness by gymnemic acid is of a mixed kinetic type which might result from two inhibition mechanisms operating simultaneously. The implication of this is that inhibition is not immediately reversible by increasing the

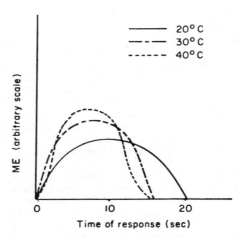

Figure 10. Time-intensity plot for sugar sweetness. Magnitude estimation (ME). Reproduced with permission from Ref. 29. Copyright 1981, Academic Press, Inc.

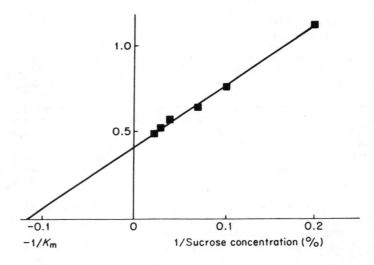

Figure 11. Reciprocal plot of MER versus concentration of sweet stimulus. Reproduced with permission from Ref. 29. Copyright 1981, Academic Press, Inc.

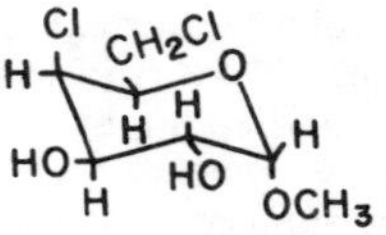

Figure 12. Structure of methyl 4,6-dichloro-4,6-dideoxy-α-D-galactopyranoside (VI), DiCl-gal.

concentration of sugar on the tongue. Thus, Birch postulates that sweetness suppression by gymnemic acid may be the result of the inhibitor binding to the receptor for a period of time (for up to an hour or so) during which time access of sugar molecules to the receptor is blocked. Also, turnover of inhibitor is much slower at the receptor than is the turnover of sugar molecules. Therefore, after a receptor site is freed of bound gymnemic acid the sugar molecules must then compete with the proximate inhibitor for access and binding.

A very powerful but short-lived inhibition of sweet taste response in the gerbil is caused by methyl 4,6-dichloro-4,6-dideoxy-α-D-galactopyranoside (DiCl-gal), VI, (Figure 12) a sugar derivative synthesized and studied by Jakinovich (26). DiCl-gal is a potent sweetness inhibitor, but only when mixed with sucrose. It has no long term inhibiting effect on the sweet taste response. Also, pretreatment of the tongue with DiCl-gal has no effect on subsequent response to sucrose solutions, in contrast to the behaviors of SDS, GA and ZJ. Kinetic analysis indicates that the substance is a competetive inhibitor, but experiments with artificial sweetners suggest that the mechanism of inhibition is possibly more complex than this.

Prospects for Future Research

There is a growing body of knowledge relating structure to function for sweet taste. In spite of this, many questions remain unanswered with regard to the physical nature of the sweetness receptor, including the question of whether it is indeed a single receptor at a single site. Once the structures of several different inhibitors are known in detail it should be possible to use them as probes of the receptors, to ascertain the physical limits and steric-electrostatic requirements of the system. By systematically varying the structure of inhibitors and measuring the psychophysical effects, structure-activity correlations can be constructed. These data, when taken together, may then allow construction of a self-consistent picture of the sweetness receptor. This, then, will allow us to better understand the taste transduction process.

Finally, it is interesting to speculate on what function sweetness inhibitors such as the gymnemic acids and the ziziphins might perform for the plants in which they occur. One thought is that they might serve as feeding deterrents for predatory insects. It has been shown that gymnemic acids act as feeding deterrents to the Southern army worn (Prodenia eridana). The effect is demonstrable with sugar-free diets, thus the gymnemic acids do not appear to act on the sweet taste in the insect, as they do in humans (27, 28). The sweetness inhibitors may also function as defenses against mammalian herbivores: by dulling the sense of sweetness, the intrinsic bitterness of a plant might be accentuated, thus encouraging the browser to find tastier fare.

Acknowledgments

Much of the ziziphin chemistry was carried out by Dr. Frank Koehn, and psychophysical studies were performed by Drs. Claire Murphy and

Carol Christensen. This work was partially supported by NIH grant 5T32NS07176-05 (postdoctoral traineeship to F.E.K.) and by the Ambrose Monell Foundation.

Literature Cited

1. Kurihara, K.; Beidler, L. M.; Science; 1968, 161, 1241-1243.
2. Brower, J. N.; van der Wel, H.; Franke, A.; Henning, G. J.; Nature (London); 1968, 220, 373-374.
3. Kurihara, K.; Beidler, L. M.; Nature (London); 1969, 222, 1176-1179.
4. van der Wel, H.; In "Olfaction and Taste, IV"; Wissenshaftliche Verlagsgesellschaft MBH, Stuttgart; 1972, 226-233.
5. Cagan, R. H.; Science; 1973, 181, 32-35.
6. DeSimone, J. A.; Heck, G. L.; Bartoshuk, L. M.; Chem. Senses; 1980, 5, 317-330.
7. Yackzan, K. S.; Ala. J. Med. Sci.; 1966, 3, 1-9, quoted in (15).
8. Subba Rao, G.; Sinsheimer, J. E.; Chem. Commun.; 1968, 1681-1682.
9. Stocklin, W.; Ag. and Fd. Chem.; 1969, 17, 704-708.
10. Sinsheimer, J. E.; Subba Rao, G.; McIlhenny, H. M.; J. Pharm. Sci.; 1970, 59, 622-628.
11. Subba Rao, G.; Sinsheimer, J. E.; J. Pharm. Sci.; 1971, 60, 190-193.
12. Pfaffmann, C.; In "Handbook of Physiology"; American Physiology Society, Washington, D. C., 1959; Section 1, Vol. 1, p. 507.
13. Dateo, G. P.; Long, L.; Agric. Fd. Chem.; 1973, 21, 899-903.
14. Stocklin, W.; Helv. Chim. Acta; 1967, 50, 491.
15. Bartoshuk, L. M.; Dateo, G. P.; Vandenbelt, D. J.; Buttrick, R. L.; Long, L.; "Olfaction and Taste, III"; Pfaffmann, C., Ed.; Rockefeller University Press, New York; 1969, pp. 436-444.
16. Meiselman, H. L.; Halpern, B. P.; Dateo, G. P.; Physiol. and Behav.; 1976, 17, 313-317.
17. Kennedy, L. M.; Halpern, B. P.; Chem. Senses; 1980, 5, 123-147.
18. Still, W. C.; Kahn, M.; Mitra, A.; J. Org. Chem.; 1978, 43, 2923-2925.
19. Ogihara, Y.; Inoue, O.; Otsuka, H.; Kawai, K.-I.; Tanimura, T.; Shibata, S.; J. Chromatogr.; 1976, 128, 218-223.
20. Kawai, K.-I.; Akiyama, T.; Ogihara, Y.; Shibata, S.; Phytochem.; 1974, 13, 2829-2832.
21. See: Cotter, R. J.; Analyt. Chem.; 1980, 52, 1589A-1606A, for review.
22. Kagawa, Y.; In "Methods in Membrane Biology"; Korn, E. D., Ed.; Plenum Press, New York, 1974; Vol. 1, pp. 201-269. Quoted in (23), p. 215.
23. DeSimone, J. A.; In "Biochemistry of Taste and Olfaction"; Cagan, R. H. and Kare, M. R., Eds.; Academic Press, New York, 1981, pp. 213-229.
24. Ray, A.; Birch, G. G.; Life Sciences; 1981, 28, 2773-2781.
25. Cagan, R. H.; Biochim. Biophys. Acta; 1971, 252, 199-206.

26. Jakinovich, W.; <u>Science</u>; 1983, 219, 408-410.
27. Granich, M. S.; Halpern, B. P.; Eisner, T.; <u>J</u>. <u>Insect</u>
    <u>Physiol</u>.; 1974, 20, 435-439.
28. Harborne, J. B.; "Introduction to Ecological Biochemistry";
    Academic Press, New York, 1977; pp. 149-150.
29. Birch, G. G.; In "Biochemistry of Taste and Olfaction";
    Cagan, R. H.; Kare, M. R., Eds.; Academic Press, New York,
    1981; pp. 163-73.

RECEIVED August 14, 1985

Sensory Responses to Oral Chemical Heat

Harry Lawless[1] and Marianne Gillette[2]

[1]S. C. Johnson & Son, Inc., Racine, WI 53403
[2]Research and Development, McCormick & Company, Inc., Hunt Valley, MD 21031

Two areas of research, psychophysics and sensory
evaluation, have made recent contributions to the
understanding of oral sensations of heat derived from
peppers. Psychophysical studies have characterized
observer's responses to heat from spice-derived
compounds, focussing on such aspects as time-intensity
functions, areas of oral stimulation, correlation with
evoked salivary flow, interactions with basic tastes,
and effects of sequential stimulation. Sensory
evaluation of the heat level of ground red pepper has
recently been advanced by the validation of a new method
which solves many of the problems inherent in the
previous Scoville procedure. The new method is based on
anchored graphic rating by panels who are trained with
physical reference standards. The procedure has shown
excellent reliability, fine discriminations among
samples, and high correlations with instrumental
determinations of capsaicinoid content of pepper samples.

Chemically-induced Oral Heat as Part of Flavor

Many varieties of red pepper, derived from plants of the genus
Capsicum, are used in different cuisines around the world for their
sensory properties of oral chemical "heat", volatile flavor and
color. Determination of the degree of heat in a pepper sample has
been a difficult problem for both sensory and instrumental analysts
of flavor. Furthermore, the literature concerning the sensory
physiology and perceptual responses of the "common chemical sense"
(as defined later) has lagged behind other areas of study of the
chemical senses. The purpose of this paper will be to review
recent developments in two areas, the development of a standard
method for sensory analysis of ground red pepper heat and the
psychophysical characterization of observers' responses to oral
chemical irritation induced by spice-derived compounds.

Physiologically, the senses responsible for our perception of flavor can be divided into three anatomical systems. In the oral cavity, the classical gustatory pathways through the tongue and soft palate are responsible for our sensitivity to the four basic tastes, sweet, sour, salty, and bitter. In the nasal passages, the olfactory receptors provide sensitivity to a wide variety of volatile compounds, producing the sensations we normally assign to smell. In addition to these two systems, the trigeminal nerves in both the oral and nasal cavities provide sensitivity to thermal, tactile, irritation and pain sensations(1). The trigeminal innervation is also chemically sensitive to compounds which are pungent, astringent or irritative, and hence provide an important part of our appreciation of flavor as a whole. This paper is concerned with the character of this system, sometimes referred to as the common chemical sense. The results presented here focus on oral chemical "heat", as characterized by the sensations (warm to painful) elicited by red or black pepper. This is distinguised from the more nasal sensation of pungency, as characterized by the naso - pharyngeal irritation induced by such substances as ammonia or freshly ground horseradish.

While many flavor compounds have irritative or astringent properties (2), most have gustatory or olfactory properties as well. However, several families of compounds, from three different spices, are potent stimuli of oral heat sensations, and in their chemically pure forms are nearly devoid of side tastes and odors. Representative members of these families are shown in Figure 1. Capsaicin is an example of the heat principles derived from red pepper. Other structures in the capsaicinoid family vary in their heat level, depending upon the saturation of the double bond (a) and the side chain length (3). N-vanillyl-n-nonamide, "synthetic capsaicin", is a readily available and easily synthesized compound used as a chemical heat standard (4, see also below), in which the branched side chain of capsaicin is replaced with a saturated straight chain. Piperine is derived from black pepper, and has three isomers depending upon the cis- or trans- configuration of the two double bonds (b and c). Gingerol is a pungent compound from ginger, with related compounds varying in pungency depending upon dehydration at (d) and upon chain length.

Two disciplines have recently brought resources to bear upon the sensory characterization of these compounds. In sensory psychology, the techniques of psychophysics have been used to characterize responses to various oral chemical irritants. Also, the field of applied sensory evaluation has addressed the issue of determining the heat value of various unknown samples using human observers. This second area of work has largely been driven by a need to replace the widespread procedure known as the "Scoville determination" for measuring sensory heat with a more reliable method. However, the orientations of the two disciplines are distinctly different. A psychophysical study will focus on determining characteristics of the observer, tend to use naive untrained subjects, simple stimuli and ask for relatively simple judgements (responses). The goal in such studies is to uncover fundamental attributes of sensory function, such as observer sensitivity, temporal and spatial properties of sensation, or dose-response curves (psychophysical functions). Sensory

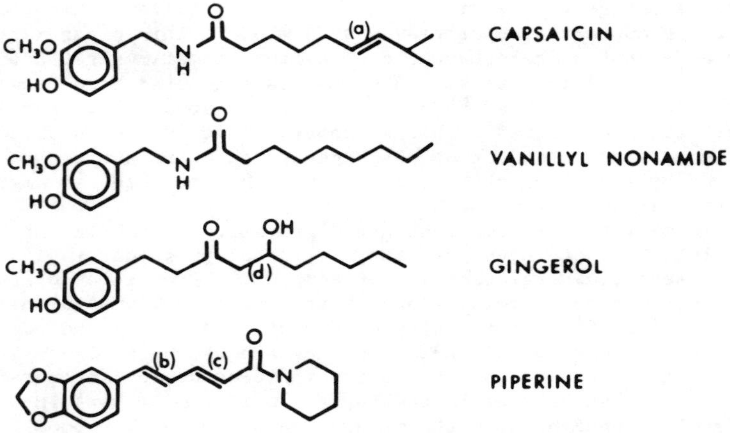

Figure 1.   Structures of flavor compounds inducing oral irritation or "heat".

evaluation, on the other hand, is oriented towards uncovering sensory attributes of the stimulus or product. This will usually be done with trained, experienced or otherwise "calibrated" observers who are used in order to elucidate the sensory properties of unknown or uncharacterized stimuli.

However, since the act of sensation is an interaction of an observer with a stimulus, the two approaches often end up using similar methods (e.g., scaling) to provide similar information (e.g., observer response as a function of stimulus concentration). Because of this parallel orientation in methods, and because of the concurrent advances recently made in both areas, we have integrated results from the two fields in this paper. The first section will focus on psychophysical characterization of oral chemical irritants. The final section will discuss the development of a new sensory method for evaluation of ground red pepper heat.

Psychophysical Characterization of Oral Irritation

One important attribute of oral chemical irritation is the long time-course of the sensations elicited, both in onset of sensation after tasting the stimulus, and in the lingering duration of the sensation after expectoration. In the results below, subjects were given emulsions of various spice-derived compounds (or mixtures) to taste. The sample was swirled vigorously around the mouth for 30 seconds, expectorated, and then various ratings were asked of the subject at fixed intervals. The method of magnitude estimation was used, in which subjects assign numbers to sensations in proportion to the sensation elicited by reference stimuli tried earlier (usually NaCl solutions). Figures 2 and 3 show data for the time-course of sensations elicited by various irritative compounds (5). To a first approximation, these functions can be characterized by the following relationship:

$$S = k \ C^n \ e^{-T/m}$$

where n is a power function exponent characterizing the growth of sensation magnitude (S) with concentration (C), and m is a constant characterizing the rate of decay of sensation over time (T) (k is a proportionality constant and e the base of the natural logarithm).

These operating characteristics, power function exponents and decay constants, can provide reference points for comparing compounds. In these studies, piperine tended to have higher exponents (faster growth of sensation with concentration) than n-vanillyl-n-nonamide. It is also worth noting that the majority of olfactory and taste compounds have exponents less than or approaching 1.0, while painful stimuli (e.g., electric shock) tend to have much higher exponents. These irritant compounds fall more in the range of flavor compounds than pain stimuli (5), with exponents less than or about equal to 1.0.

Another characteristic of irritative stimulation of the trigeminal nerve is the defensive reflexes (e.g., sneezing) invoked by the body to remove or dilute the offending substance. In the case of oral chemical heat, the burning sensation from capsaicin invokes sweating, tearing, and copious salivation. Salivary flow

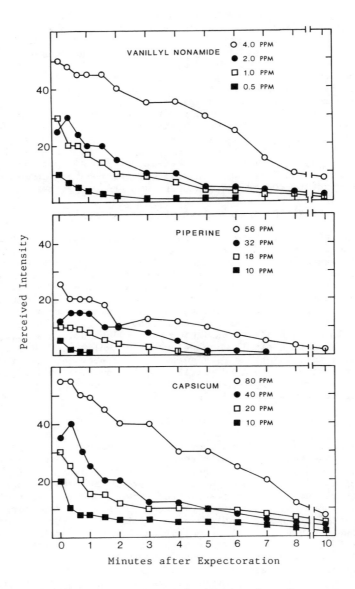

Figure 2. Median perceived oral irritation from four concentrations of vanillyl nonamide, piperine and capsicum oleoresin over time. Reproduced with permission from Ref. 5, copyright 1984, IRL Press Limited.

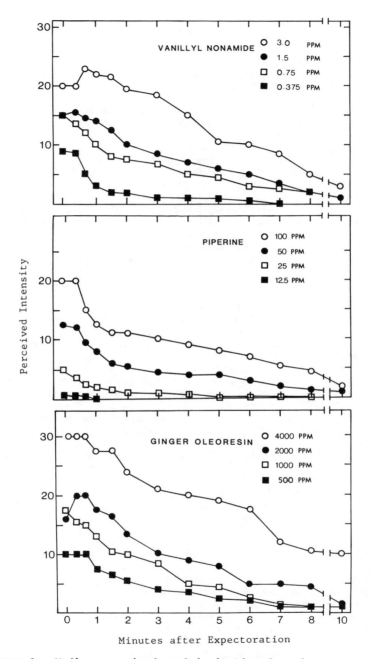

Figure 3.   Median perceived oral irritation from four
concentrations of vanillyl nonamide, piperine and ginger
oleoresin over time.   Reproduced with permission from Ref. 5,
copyright 1984, IRL Press Limited.

rate closely parallels the subjective ratings of sensation
intensity. Whole-mouth salivary flow was collected during the
initial two minutes of oral burn from the subjects who tasted the
twelve compounds in Figure 2. The correlation of mean salivary
flow rate with mean peak sensation intensity was .91. The
"subjective" ratings, in this case, find a source of validation
since a high correlation exists with an "objectively" measured
physiological response.

In addition to the pattern of stimulation over time, these
compounds also show similarities and differences in spatial
patterns of stimulation. After a whole-mouth rinse, subjects were
asked to report the areas of the mouth in which they perceived the
heat. Not surprisingly, the number of oral areas that subjects
reported systematically increased with concentration, and decreased
over time as the burn faded in intensity (5). The patterns of
perceived stimulation across the mouth also differed among
compounds. The red pepper compounds and n-vanillyl-n-nonamide
showed pronounced anterior stimulation (areas of lips, gums,
tongue). The ginger compounds, on the other hand, also showed some
posterior stimulation, with many subjects reporting a biting
sensation on the soft palate and throat. These data suggest that
compounds may differ qualitatively in the areas they most
efficiently stimulate.

One further aspect in which irritative compounds differ is
their interaction with the tastes, sweet, sour, salty and bitter.
A recent perceptual study sought to gain insights on the old adage
that too much pepper makes it hard to taste your food. In this
study, stimuli representing the classical four basic tastes were
given after rinses with capsicum oleoresin (an extract of red
pepper) or piperine. Psychophysical ratings of the perceived
intensity of the tastants showed that under conditions of intense
oral irritation, there was some partial inhibition of the taste
sensations (6). In addition, the pattern of inhibition differed
for the two irritants. The pungency of capsicum worked mostly
against sour and bitter tastes, and left saltiness intact, while
the effects of piperine were more broad, influencing all four taste
qualities. These inhibitory effects of irritation on taste are
parallel to similar effects of nasal irritation on odors. For
example, CO_2, a potent nasal irritant, will partially mask odors
which are presented simultaneously (8). Whether oral burn can
influence odor or aroma perception is at this time unknown.

Since a series of several stimuli are usually presented to
subjects in psychophysical studies, the opportunity arises to study
effects of sequential stimulation. These effects can have
practical implications such as the number of stimuli which may be
sampled in applied sensory evaluation without fatigue. During the
time-intensity ratings which produced the data in Figures 2 and 3,
stimuli were presented in different orders, with the first and
third concentrations in the series of each compound presented
either before or after the stimulus of next higher concentration.
As shown in Figure 4, stimuli were usually judged to be much weaker
when presented after a stronger compound. This suggests that there
is a sequential desensitizing effect during tasting and that
experimenters should be careful concerning the number of stimuli
that may be reasonably given in one sitting. This desensitization

parallels the effects of capsaicin observed in the pharmacological literature (8,9), where systemic or topical administration renders animals insensitive to chemical irritation. Whether or not people became desensitized during long-term dietary intake of pepper is unclear. One study examining thresholds failed to see a difference between chili consumers and non-consumers (10). However, lack of effects at threshold may not reflect above-threshold changes in responsiveness.

However, short-term desensitization may depend upon the spatial characteristics of stimulation. When stimuli are constantly refreshed, and stimulation is limited to a small area, a pattern of ever-increasing sensation buildup is observed. Figure 5 shows a flow chamber ("geschmackeslupe" after Hahn (11)) used to deliver constant controlled stimulation. When vanillyl nonamide was flowed over a subject's tongue (2 ml/sec, 0.8 cm^2 area, 8 subjects), the sensation continued to grow as shown in Figure 6, about doubling during a period of seven minutes. Further psychophysical characterization may better delineate the parameters that influence these sensations and their spatial and temporal interactions.

<u>Sensory Evaluation of Pepper Heat</u>

Commercially, ground red peppers are purchased, sold, blended, and used based upon their sensory heat levels. Generally, the higher the heat, the higher the price. In order to produce a consistent product the heat level of capsicum products are monitored by sensory and chemical/instrumental methods (12-15). Historically, the only sensory method for the assessment of heat in red pepper has been the Scoville Heat Test (16, 17). While this method originally filled the need for a means of measuring and expressing heat in red pepper products, it has become universally criticized for its lack of accuracy and precision (4, 15, 18-20). Specific problems noted with the Scoville Heat Test are: build up of heat, rapid taste fatigue and increased taste threshold as a result of the 5 samples required for tasting, ethanol bite interfering with capsicum heat, lack of statistical validity, lack of reference standards, the 16 hour extraction time, the error of central tendency (tendency to pick the middle concentration of the series) and poor precision.

The development of new instrumental methods to replace the troubled Scoville procedure necessitated the design of an improved sensory method for validation of instrumental precision. A new sensory method (4) was designed which offered the following procedural improvements over the Scoville Method: 1) a 20 minute aqueous extraction, (2) no ethanol used, 3) reference standard included in each test, 4) trained panelists, 5) timed tasting, rinsing and recess, 6) one dilution for all samples, and 7) use of a graphic line scale to score the heat sensation.

To evaluate the aqueous extraction procedure, ground red pepper was extracted with spring or distilled water, at 20°C, 75°C, and 90°C, with or without 20-2000 mg/L Polysorbate - 80 or Polysorbate - 60. The aqueous extractions were compared to 5 hour ethanol extractions. Residues from all extractions were also evaluated for residual heat. A 20 minute simmering extraction of

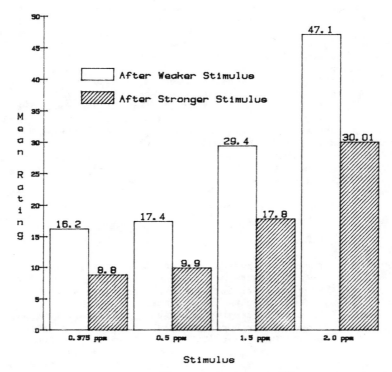

Figure 4. Mean perceived intensity (at the peak of the
time-intensity function) of vanillyl nonamide in different
presentation orders.

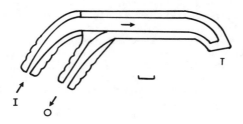

Figure 5. Geschmackeslupe for presenting flowing stimuli to
oral surfaces. Scale = 1 cm. I - inflow port, O - outflow
port, T - port for stimulation, approximately 1 cm in
diameter. Inflow and outflow tubes in the barrel are
concentric.

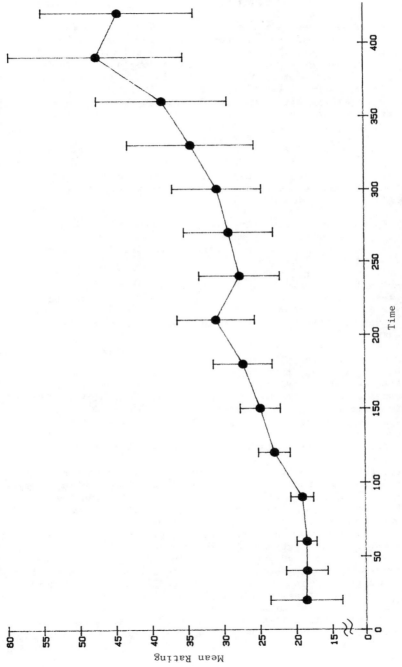

Figure 6. Mean ($^+$ 1 S.E.m) perceived intensity of 2 ppm vanillyl nonamide during presentation through Geschmakeslupe, over seven minutes.

ground red pepper in 90°C spring or distilled water, with 200 ppm
polysorbate –60 or –80 was found to optimize extraction of the
sensory heat components.
    A panel of 12 highly experienced tasters was used to develop
the new method. Pepper heat was rated on a 15 cm line scale
anchored at 0 (no heat), 1.25 cm (threshold heat), 5.0 cm (slight
heat), 10.0 cm (moderate heat), and 15 cm (strong heat). The panel
selected standard concentrations of N–vanillyl–n–nonamide to be
used for calibration and training of future panelists. A reference
control (0.44 ppm N–vanillyl–n–nonamide) was selected for use an an
internal standard for "slight" during each sensory test. Red
lights were used to eliminate possible influences of variation in
the color of the products. Two samples were evaluated per test,
the known control and a test pepper extract. The control, coded
"C" was served first, followed by an unknown test sample identified
with a random double letter code. All samples were presented as 10
ml portions in plastic medicine cups. Panelists evaluated all
samples using the following procedure:

1. Cleanse palate before first sample (control) with unsalted
    cracker and 20°C spring water.
2. Take entire first sample (control) in mouth, hold for about
    5 seconds, swallow slowly.
3. Wait 30 seconds (timed).
4. Rate first sample at "slight" on ballot.
5. Cleanse palate with unsalted cracker and 20°C spring water
    for 60 seconds (timed).
6. Rinse with 20°C spring water immediately prior to second
    sample.
7. Take entire second sample (test sample) in mouth for about 5
    seconds, swallow slowly.
8. Wait 30 seconds (timed).
9. Rate second sample.
10. Panel dismissed if only one sample is to be evaluated. If
    two samples are to be evaluated:
11. Wait 5.0 minutes. Clease palate well with water and
    crackers during this time.
12. Repeat steps 1 through 9 for the second set of samples.

Panelists placed a mark on the scale expressing their impression of
the heat in the test sample. Sensory Heat Ratings were obtained by
measuring the distance in cm from the "0" mark to the panelist's
rating for each sample. The mean of all panelist's ratings for
each sample represents its sensory heat rating.
    To evaluate the correlation of sensory responses with High
Pressure Liquid Chromatography capsaicinoid quantitation, samples
from 60 lots of ground red pepper were selected to represent the
normal range of Scoville Heat Units found in red pepper. These 60
peppers were analyzed instrumentally for the 3 major capsaicinoid
analogs (nordihydrocapsaicin, capsaicin and dihydrocapsaicin), for
8 additional physical/chemical parameters (water activity,
moisture, color, surface area and particle size), and sensorially
by the Scoville heat tests. The 60 peppers were also assessed
using the new sensory method for heat ratings. All possible single

and multiple regressions were performed in order to determine the
optimal instrumental alternative for the new sensory method, as
well as to further substantiate the precision of the new sensory
method (4). Several very strong relationships (r ≥ 0.90) were
found between the instrumental and sensory measurements (Table 1).

   The strong correlation (r=0.94) between the total concentration
of capsaicinoids and sensory heat (Table I), and the relative ease
of measuring such, lead to the selection of this chemical parameter
as the most desirable predictor of the sensory heat. The
predictive strength of the equation relating percent capsaicinoids
with sensory heat ratings was then tested using 20 red peppers.
The mean average deviation between predicted and actual sensory
heat ratings was less then 1 cm on the 15 cm scale (4). Thus, the
equation "sensory heat rating = 31.26 (percent capsaicinoids) –
0.21" precisely predicts sensory ratings for heat in ground red
pepper.

   Lesser predictive relationships were found between the
traditionally determined Scoville heat units and either the sensory
heat ratings (r=0.40) or the percent capsaicinoids (r=0.48). Based
upon the sensory heat values for individual capsaicinoids as
determined by Todd, et. al. (15), Scoville – type heat units were
calculated for each red pepper based upon the capsaicinoid content
of the peppers (13). Linear regression demonstrated the
relationship between these calculated Scoville heat units and the
new sensory heat ratings to be very good (r=0.94) (Figure 7).
Therefore, both the HPLC and the new sensory method can provide
output translated into Scoville units for universal understanding.
The accuracy of the new sensory method was confirmed using a set of
15 "artificial" red peppers of known oleoresin capsicum content
(Figure 8). The correlation of sensory heat rating with the
percent oleocapsicum was 0.94. The precision of the new method was
demonstrated by collaborative study within 13 laboratories (22),
and by repeated testing within one lab (4) (Tables II and III).

   Furthermore, the new sensory method avoids several problems
inherent in the Scoville procedure. Heat build up, fatigue, and
increased threshold are minimized by use of a standardized initial
sample, as well as timed rinsing between samples. Ethanol bite is
avoided by use of an aqueous extraction. The panel data may be
manipulated statistically due to the linearity of the scale and the
number of panelists. Reference standards are included. Extraction
time is reduced from 16 hours to 20 minutes. Reproducibility of
results has been demonstrated. The error of central tendency is
avoided by not having a "middle" sample. The new method is more
comparable to normal food usage as it is an aqueous rather than
ethanol extraction.

   This method is currently being used for routine laboratory
analysis of red pepper heat. Results have been consistent and
continue to correlate well with HPLC data. A similar procedure has
also been used for sensory evaluation of black pepper heat. The
American Society for Testing and Materials (ASTM, Committee E–18)
has conducted a collaborative study testing the new method in
comparison to the Scoville Method. ASTM E–18 is currently
preparing to document it as a standardized test method. Also, a
modification of the method is being prepared for oleoresin capsicum
and for low–heat capsicums.

Table I. Results of Regression Analyses on Sensory Heat Versus Several Analytical Measurements of 40 Ground Red Peppers

Variables	r	F	Equation of line
Sensory vs:			
% Capsaicinoids[a] and ΔE[b]	0.959	212.6**	Sensory = 34.18 (% Capsaicoinoids) − 0.216 (ΔE) + 10.24
% Capsaicinoids and b[c]	0.954	188.8**	Sensory = 34.66 (% Capsaicinoids) − 0.364 (b) + 7.07
% Capsaicinoids and a_w[d]	0.946	159.0**	Sensory = 30.38 (% Capsaicinoids) + 0.522 (Aw) − 2.12
% Capsaicinoids and Moisture[e]	0.943	151.1**	Sensory = 31.02 (% Capsaicinoids) + 10.272 (Moisture) − 1.61
% Capsaicinoids	0.939	284.1**	Sensory = 31.26 (% Capsaicinoids) − 0.214
Calaulated Scoville Heat Units[f]	0.938	280.5**	Sensory = 1.99 (Calculated Scoville) − 0.223
% Capsaicin[g]	0.921	214.9**	Sensory = 45.54 (% Capsaicin) + 0.256
% Nordihydrocapsaicin[h] and ΔE	0.899	78.3**	Sensory = 468.32 (% Nordihydrocapsaicin) − 0.149 (ΔE) + 8.39
Scoville Heat Units[i]	0.475	11.1**	Sensory = 1.45 (SHU) + 0.303
Scoville Heat Units vs:			
Calculated Scoville Heat Units	0.476	5.5	Calculated Scoville = 0.683 (SHU) + 4684.67
Sensory	0.475	11.1**	Sensory = 1.45 (SHU) + 0.303

[a] Total capsaicinoids as determined by HPLC (Hoffman et al., 1983).
[b] Change in total color, determined by Hunter Colorimeter, Model D25M-9.
[c] Color value, Hunter Colorimeter, Model D25M-9.
[d] Water activity determined by Beckman Water Activity Hygrometer.
[e] Moisture determined by Azeotropic Distillation.
[f] Calculated using method described by Todd et al. (1977).
[g] Capsaicin as determined by HPLC (Hoffman et al., 1983).
[h] Nordihydrocapsaicin as determined by HPLC (Hoffman et al., 1983).
[i] ASTA method 21.0; Scoville Heat Test.
** Statistically significant at 99% level of confidence.

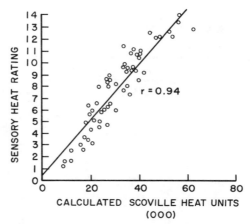

Figure 7.   Sensory test heat ratings versus calculated Scoville
Heat units (in thousands) for 60 red peppers.   Scoville Heat
units calculated based upon the method of Todd, et.al. (15).
Reproduced with permission from Ref. 4, copyright 1984,
Institute of Food Technologists.

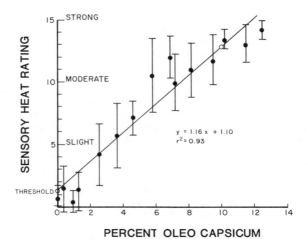

PERCENT OLEO CAPSICUM

Figure 8.   Sensory heat ratings versus concentration of
oleoresin capsicum on paprika for a set of 15 artificial red
peppers.   Reproduced with permission from Ref. 4, copyright
1984, Institute of Food Technologists.

Table II.   Sensory Heat Ratings for Blind Duplicate
            Samples of Red Pepper[a]

Pepper	Sensory heat rating Session 1	Sensory heat rating Session 2
A	3.3	2.7
B	3.5	4.6
C	5.7	5.9
D	5.9	5.6
E	6.6	4.8
F	8.5	9.8
G	8.5	8.6
H	8.6	8.8
I	10.7	10.9
J	10.9	10.9

[a] Each sample evaluated at 2 different sessions by the same panel
(n = 10). No pair of duplicate samples is different when analyzed
by a paried t test.

CONCLUSIONS

1. Sensations of oral chemical heat, as induced by pepper-derived
    compounds are amenable to psychophysical investigation and
    sensory evaluation by supra-threshold scaling techniques such
    as magnitude estimation and anchored graphic rating scales.

2. Different heat-inducing compounds may be characterized by
    different pschophysical functions, perceived areas of oral
    stimulation, and interactions with taste sensations.

3. Repeated sampling of intense oral heat stimuli may result in
    short-term desensitization. However, this effect may depend
    upon the spatial and temporal parameters of stimulation, since
    constant stimulation of small areas of the oral epithelium
    leads to sensation growth, rather than desensitization.

4. A new rating method for evaluation of the sensory heat of
    ground red pepper samples shows important advantages over the
    traditional Scoville method, especially in the areas of
    accuracy, reliability and ease of administration.

5. The new sensory method shows excellent correlation with
    instrumental determination of capsaicinoid content of red
    pepper samples, and can be converted to Scoville units for
    universal understanding.

Table III. Results of Collaborative Study on Red Pepper Sensory Heat Method[a]

Sensory heat ratings

Pepper	Lab #1			Lab #2	Lab #3	Inter-Lab X (n = 5)	Inter-Lab σ (n = 5)	Inter-Lab σ/X (n = 5)
	Panel A (n = 10)	Panel B (n = 10)	Panel C (n = 9)	Panel D (n = 9-11)	Panel E (n = 4-5)			
1	0.6 ± 0.6	0.6 ± 0.6	0.4 ± 0.5	0.5 ± 0.8	1.0 ± 1.1	0.6	0.23	0.38
2	3.0 ± 1.7	4.0 ± 2.3	4.7 ± 2.3	4.8 ± 1.3	6.1 ± 1.4	4.5	1.1	0.22
3	5.4 ± 2.1	5.1 ± 3.1	3.4 ± 2.0	6.0 ± 2.0	6.5 ± 1.4	5.3	1.2	0.22
4	8.4 ± 2.5	9.4 ± 2.1	8.3 ± 1.7	4.4 ± 1.7	7.5 ± 1.4	7.6	1.9	0.25
5	10.0 ± 1.4	7.8 ± 2.4	10.0 ± 1.7	9.7 ± 2.1	11.0 ± 1.4	9.7	1.2	0.12
6	12.6 ± 1.7	9.1 ± 1.7	11.3 ± 0.9	11.9 ± 1.7	12.0 ± 1.4	11.4	1.4	0.12

[a] Means and standard deviations for 6 ground red peppers tested in 3 labs by 5 separate panels. Laboratory Means, Standard Deviations and Coefficients of Variations. Coefficient of Variation = σ/X; an approximation-of method inter-laboratory precision.

Literature Cited

1. Silver, W. L.; Maruniak, J. A.; Chem. Senses 1981, 6, 295-305.
2. "Fennaroli's Handbook of Flavor Ingredients," Second Edition; Furia, T. H. and Bellanca, N., Ed.; Cleveland, 1975; Vol. II.
3. Govindarajan, V. S. In "Food Taste Chemistry"; Boudeau, J. C., Eds.; ACS SYMPOSIUM SERIES No. 115, American Chemical Society; Washington, D.C, 1979; p. 53.
4. Gillette, M. H.; Appel, C. E.; Lego, M.C. Journal of Food Science 1984, 49.
5. Lawless, H.; Chem. Senses 1984, 9, 143-155.
6. Lawless, H.; Stevens, D. A.; Physiol. Behav. 1984, 32, 995-998.
7. Cain, W. S.; Murphy, C. L.; Nature 1980, 284, 255-257.
8. Jancso, N.; Bull Millard Fillmore Hosp. Buffalo 1960, 7, 53-57.
9. Nagy, J. I.; "Handbook of Psychopharmacology," New York, Vol. 15
10. Rozin, P.; Schiller, D.; Mot. Emot. 1980, 4, 77-101.
11. Hahn, H.; "Beitrage fur Reizphysiologie", Heidelberg, 1949.
12. Bajaj, K. J. AOAC. 1980. 63(6):1314.
13. Hoffman, P.G.; Salb, M.C.; Galetto, W.G. J. Agr. Food Chem. 1983. 31(6:1326).
14. Palacio, J. J. AOCA. 1977. 60(4):970
15. Todd, P.H.: Bensinger, M.G.: Biftu, T. J. Food Sci. 1977. 42(3):660.
16. Scoville, W.L. J. Amer. Pharm. Assn. 1912. 1 : 453.
17. American Spice Trade Association. 1968. "Official Analytical Methods" Method 21.0.
18. Suzuki, J.I.; Tasign, F.; Morse, R.E. Food Technol. 1957. 11:100.
19. Maga, J.M. Critical Rev. Food Sci Nutrition. 1975. July : 177.
20. Govindajaran, V.S.; Narasimhan S.; Khanaraj, S.; JFS&T, India, 1977, 14(1):23.
21. Rhyu, H.Y. J. Food Sci. 1978. 43(5): 1632
22. American Society for Testing and Materials. Committee E-18 on Sensory Evaluation, Subcommittee E-18, 03 on "Other Senses". Unpublished data.

RECEIVED August 5, 1985

Analysis of Chiral Aroma Components in Trace Amounts

R. Tressl, K.-H. Engel, W. Albrecht, and H. Bille-Abdullah

Forschungsinstitut für Chemisch-technische Analyse, Technische Universität Berlin, Seestrasse 13, D-1000 Berlin 65, Federal Republic of Germany

Methods for the capillary gas chromatographic separation of optical isomers of chiral compounds after formation of diastereoisomeric derivatives were developed. Analytical aspects of the GC-separation of diastereoisomeric esters and urethanes derived from chiral secondary alcohols, 2-, 3-, 4- and 5-hydroxy-acid esters, and the corresponding γ- and δ-lactones were investigated. The methods were used to follow the formation of optically active compounds during microbiological processes, such as reduction of keto-precursors and asymmetric hydrolysis of racemic acetates on a micro-scale. The enantiomeric composition of chiral aroma constituents in tropical fruits, such as passion fruit, mango and pineapple, was determined and possible pathways for their biosynthesis were formulated.

Capillary column gas chromatography has proved to be a suitable technique for the determination of the enantiomeric composition of chiral compounds in trace amounts. The gas chromatographic resolution of chiral substances can be achieved either by separation of enantiomers on an optically active stationary phase (1) or by formation of diastereoisomeric derivatives and analysis on a non-chiral phase (2).
Separation of enantiomers on chiral phases without derivatizations have been described only for a few compounds, such as monoterpenes (3). Most separations require derivatization of the enantiomers, e.g. conversion into N-containing derivatives (4), and at present the number of available thermally-stable chiral stationary phases is limited.
During our studies of the biogenesis of chiral compounds, we developed micro-methods for the analysis of

0097-6156/85/0289-0043$06.00/0

naturally occuring trace-components using non-chiral phases. In this paper we describe two methods for forming
diastereoisomeric derivatives suitable for gas chromatographic investigations of a broad spectrum of chiral
hydroxy compounds, the formation of diastereoisomeric
esters of R-(+)-α-methoxy-α-trifluoromethylphenylacetic
acid (R-(+)-MTPA), and of distereoisomeric urethanes of
R-(+)-phenylethylisocyanate (R-(+)-PEIC).
     By using high resolution capillary gas chromatogra-
phy, the scope of use of these reagents, which were ini-
tially introduced for NMR-analysis (5) and for GC separation of secondary alcohols (6), was enlarged considerably
(7,8). The application of these methods to the analysis
of optically active compounds formed during microbiological processes and to the determination of the enantiomeric composition of chiral aroma constituents in some
tropical fruits is described in this paper.

Analytical aspects of the GC-separation of diastereoisomeric R-(+)-MTPA and R-(+)-PEIC derivatives

Alcohols

     In Table I the data for the capillary GC-separation
of diastereoisomeric R-(+)-MTPA esters and R-(+)-PEIC
derivatives of some chiral aliphatic alcohols are summarized. For comparison, the data for the separation of
isopropylurethane derivatives of alcohol enantiomers on
an optically active stationary phase (9) are also listed.
     It is obvious that the separations are strongly de-
pendent on the structures of the alcohols. The highest
separation factors within the shortest time of analysis
were obtained with diastereoisomeric urethane derivatives.
     The methods described are also useful for the in-
vestigation of chiral terpene alcohols. Figure 1 presents
the separation of the eight menthol stereoisomers. Neomenthol (1), neoisomenthol (2), menthol (3) and isomen-
thol (4) can be separated and isolated by preparative GC.
By formation of R-(+)-MTPA esters, only neomenthol and
menthol could be resolved. In contrast, menthol stereoisomers could be separated after derivatization with
R-(+)-PEIC. Schurig and Weber (3) separated the enantiomers of menthol by complexation capillary gas chromatography. Benecke and König (10) investigated isopropylurethane derivatives using a chiral stationary phase.

Table I. Separation Data for the Capillary GC-Separation of Chiral Secondary Alcohols

Compound	R-(+)-MTPA(a)			R-(+)-PEIC(b)			IPIC(c)		
	T(°C)	t_{R1} t_{R2}	α	T(°C)	t_{R1} t_{R2}	α	T(°C)	t_{R1} t_{R2}	α
Pentanol-2	170	6.53 6.65	1.018	160	6.45 6.70	1.039	115	5.60	–
Heptanol-2	170	14.50 14.80	1.021	160	12.30 13.13	1.067	115	19.00 19.30	1.016
Octanol-2	170	22.45 22.98	1.024	160	17.33 18.75	1.082	115	32.92 33.40	1.015
Nonanol-2	170	34.38 35.13	1.022	175	12.90 11.90	1.084	115	48.70 49.50	1.016
Undecanol-2	180	28.23 28.63	1.014	185	15.00 16.38	1.092	130	65.85 66.95	1.017
Octanol-3	170	20.80 21.05	1.012	160	15.38 16.22	1.055	115	27.70 28.30	1.022
Octanol-4	170	18.05	–	160	14.20 14.45	1.018	115	21.55	–
1-Octen-3-ol	170	19.25 20.05	1.042	160	15.38 16.22	1.055	115	32.50	–

(a) separation on a glass capillary column OV 101; 50 m/0.32 mm i.d.

(b) separation on a fused silica column DB 210; 30 m/0.33 mm i.d.

(c) separation of the isopropylisocyanate (IPIC) derivatives on a fused silica column XE-60-L-valine-(S)- -phenylethylamid, 50 m/0.25 mm i.d.

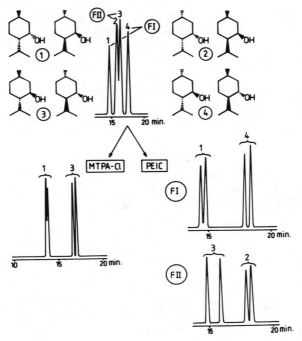

Figure 1. Capillary GC separation of menthol isomers as R-(+)-MTPA derivatives (CP Wax 57 CB, 50 m/0.32 mm i.d., 180 °C) and R-(+)-PEIC derivatives (F I: CP Wax 57 CB, 50 m/0.32 mm i.d., 230 °C; F II: DB 210, 30 m/0.33 mm i.d., 180 °C).

2- and 3-Hydroxyacid esters

Table II presents the data for the GC-separation of
R-(+)-MTPA and R-(+)-PEIC derivatives of 2- and 3-hydro-
xyacid esters.

Whereas R-(+)-MTPA esters of 2-hydroxyacids can be
resolved on packed columns (11), the separation of R-(+)-
MTPA derivatives of chiral 3-hydroxyacids requires an
efficient capillary column. The R-(+)-MTPA-esters possess
higher separation factors than the PEIC derivatives, which
depend on the chain length of the acid. The separation of
R-(+)-MTPA derivatives of 3-hydroxyacid esters is influenced by the stationary phase. The efficiency of separation
improves with increasing polarity of the phase, the
highest α-value was obtained with a bonded trifluoro-
propylmethylsilicone phase (12).

Table II. Separation Data for the Capillary GC-
Separation of R-(+)-MTPA and R-(+)-PEIC Derivatives
of Some 2- and 3-Hydroxyacid Esters

Compound	R-(+)-MTPA [a]			R-(+)-PEIC [a]		
	$T(^{\circ}C)$	t_{R1} t_{R2}	α	$T(^{\circ}C)$	t_{R1} t_{R2}	α
Ethyl 2-hydroxy-propionate	170	5.50 6.30	1.134	190	6.05 6.30	1.041
Ethyl 2-hydroxy-butanoate	170	6.60 7.75	1.174	190	7.63 7.85	1.030
Ethyl 2-hydroxy-hexanoate	190	5.10 5.90	1.157	190	12.43 12.60	1.014
Ethyl 3-hydroxy-butanoate	180	7.30 7.68	1.052	210	4.85 5.05	1.041
Ethyl 3-hydroxy-pentanoate	180	8.80 9.20	1.045	200	7.65 7.95	1.039
Ethyl 3-hydroxy-hexanoate	190	8.65 9.10	1.052	200	9.55 9.80	1.026

(a) column: DB 210, 30 m/0.33 mm i.d.

4- and 5-Hydroxyacid esters

The gas chromatographic separation of enantiomers
or diastereoisomeric derivatives of 4-hydroxyacids and

the corresponding chiral γ-lactones, respectively, has
not been reported previously. Outlined in Figure 2 are
the derivatization procedures that we used to obtain diastereoisomeric derivatives of chiral γ-lactones, which
could be separated by capillary GC.

Table III presents separation data for R-(+)-MTPA
derivatives of some 4-hydroxyacid esters obtained from
chiral γ-lactones. The esterified alcoholic substituent
strongly influenced the separation. Since the α-values
for the separation of R-(+)-MTPA derivatives of methyl-
esters were very low, the lactones were converted to
4-hydroxyacid ethylesters by interesterification with so-
diumethylate. Although base line separation was not at-
tained (using a 30 m column), the method could be applied
to control the formation of optically pure γ-lactones
and 4-hydroxyacid esters during microbiological processes
(13).

Table III. Separation Data for the Capillary GC-
Separation of R-(+)-MTPA Derivatives of Some
4-Hydroxyacid Esters

R-(+)-MTPA [a] derivatives	T($^{\circ}$C)	t_{R1} t_{R2}	$\dfrac{t_{R2}}{t_{R1}}$
Methyl 4-hydroxy-hexanoate	170	16.00 16.15	1.009
Methyl 4-hydroxy-octanoate	175	20.45 20.58	1.006
Ethyl 4-hydroxy-pentanoate	175	13.30 13.53	1.017
Ethyl 4-hydroxy-hexanoate	180	12.40 12.65	1.020
Ethyl 4-hydroxy-octanoate	185	15.40 15.60	1.013

(a) column: DB 210; 30 m/0.33 mm i.d.

The best separation of chiral γ-lactones was reached after derivatization of 4-hydroxyacid esters with
R-(+)-phenylethylisocyanate. The α-values for the separation of γ-hexalactone are given in Table IV. Unfortunately, the α-values decrease drastically with increasing
chain length, so that 4-hydroxyoctanoic acid and the higher homologues could not be separated by this method.

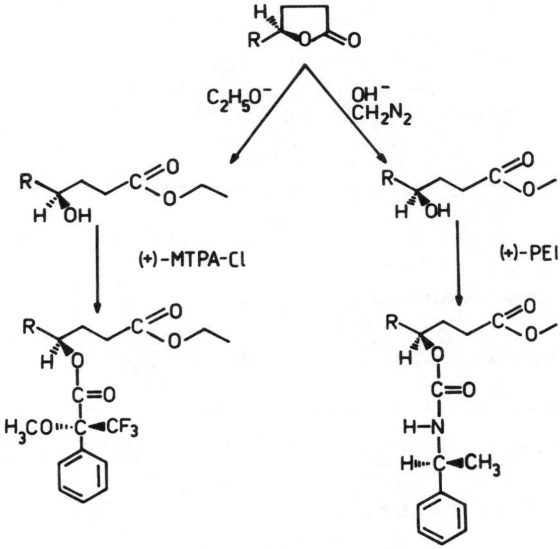

Figure 2. Conversion of γ-lactones to R-(+)-MTPA- and R-(+)-PEIC-derivatives of the corresponding 4-hydroxy acid esters.

Whereas R-(+)-MTPA-derivatives of 5-hydroxyacid
esters, obtained from chiral δ-lactones, could not be
separated by capillary GC, the corresponding R-(+)-PEIC
derivates showed excellent α-values. The separation factors for R-(+)-phenylethylurethanes, derived from some
chiral δ-lactones are listed in Table IV. The α-values
also depend on the alcoholic substituent and decrease
with increasing chain length of the lactones.

Table IV. Separation Data for the Capillary GC-
Separation of R-(+)-Phenylethylurethanes of 4-
and 5-Hydroxyacid Esters Derived from Chiral
γ- and δ-Lactones

Compound [(a)]	T (°C)	t_{R1} t_{R2}	$\alpha = \frac{t_{R2}}{t_{R1}}$
Methyl 4-hydroxyhexanoate	200	8.97 9.32	1.039
Ethyl 4-hydroxyhexanoate	200	10.50 11.05	1.052
Methyl 5-hydroxyhexanoate	200	10.40 11.05	1.063
Ethyl 5-hydroxyhexanoate	200	12.30 13.20	1.073
Methyl 5-hydroxyoctanoate	210	9.43 9.73	1.032
Methyl 5-hydroxynonanoate	210	11.88 12.13	1.021
Methyl 5-hydroxydecanoate	210	14.95 15.10	1.010

(a) column: DB 210; 30 m/0.33 mm i.d.

Investigation of chiral compounds formed during micro-biological processes

Optically pure compounds, which are obtained during
microbiological processes, can be used to determine the
order of elution of gas chromatographically separated
diastereoisomers. On the other hand the formation of
chiral compounds during microbiological processes can be
followed by capillary GC investigation of diastereoisomeric derivatives on micro-scale without working up large
sample amounts.

The reduction of methylketones to the corresponding
S-(+)-alkan-2-ols by actively fermenting baker's yeast
(14,15) was used to determine the GC order of elution of
R-(+)-MTPA esters of secondary alcohols (Figure 3). The
asymmetric hydrolysis of racemic acetates by microorga-
nisms is a further method to obtain optically pure alcohols (16). Capillary GC investigation of the R-(+)-MTPA
derivatives revealed, that enzymic hydrolysis of racemic
2-heptyl acetate by Candida utilis led to R-(-)-hepta-
nol-2 (Figure 3). A similar result was obtained for the
hydrolysis of racemic 2-octyl acetate by Candida utilis
or Brevibacterium ammoniagenes, which led to R-(-)-octa-
nol-2 (93 % R-(-)). It seems to be a common principle,
that alcohol enantiomers obtained by NADH-dependent ketone reduction are opposite to those obtained by asymmetric hydrolysis of racemic acetates.
      A similar effect could be observed when we investi-
gated the reduction of 3-ketoacid esters by Saccharomy-
ces cerevisiae and the hydrolysis of 3-acetoxyacid esters
by Candida utilis (17). As shown in Figure 4 the chiral
3-hydroxyacid esters obtained by these two processes possessed opposite configurations.
      The configuration of 3-hydroxyacid esters obtained
by yeast reduction of 3-ketoprecursors depends on the
chain length of the acids (18,19). Zhou et al. (20) demonstrated that the structure of the alcohol esterified
in 3-ketoacid esters also influences the configuration of
the 3-hydroxycompounds that are formed. The fact, that
the optical purity of the products depends on the concentration of the 3-ketocompounds (21) indicates the presence of competing enzymes leading to opposite enantiomers at different rates.
      The hydrolysis of racemic 3-acetoxyacid esters by
Candida utilis was carried out in micro-sale and the configuration of the obtained 3-hydroxycompounds was determined by capillary GC investigation of R-(+)-MTPA derivatives. As shown in Figure 4, the stereochemical course of
the hydrolysis also depends on the chain length of the
acids, and the presence of several yeast hydrolases leading to opposite enantiomers of 3-hydroxyacid esters
seems probable.
      The stereochemical course of the asymmetric hydroly-
sis of racemic 2-acetoxyacid esters by Candida utilis (17)
is demonstrated in Figure 5. The yield of the obtained
chiral 2-hydroxyacid esters was very low in comparison
with the hydrolysis of the corresponding 3-acetoxycom-
pounds. The determination of the optical purity was only
possible by isolating the products by preparative GC and
by capillary GC investigation of their R-(+)-MTPA derivatives. As shown in Figure 5, the configuration of the
isolated 2-hydroxyacid esters changed from (S) to (R)
with increasing chain length of the acids. There is a
preverential hydrolysis of the (S)-2-acetoxyacid ester
to give the corresponding (S)-configurated hydroxycom-

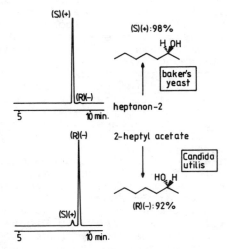

Figure 3. Formation of optically pure alcohols by
stereospecific reduction of ketones and asymmetric hydrolysis of racemic acetates (DB 210, 30 m/0.33 mm i.d.).

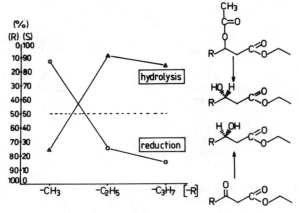

Figure 4. Enantiomeric composition of 3-hydroxyacid
esters formed by reduction of 3-ketoacid esters (baker's
yeast) and hydrolysis of 3-acetoxyacid esters (Candida
utilis).

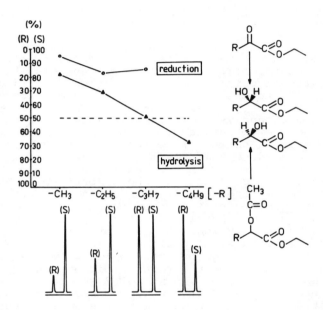

Figure 5. Enantiomeric composition of 2-hydroxyacid esters formed by reduction of 2-ketoacid esters (baker's yeast) and hydrolysis of 2-acetoxyacid esters (<u>Candida</u> <u>utilis</u>).

pound. With increasing chain length, however, there also
seems to be a preferential metabolism of this enantiomer,
which finally leads to the (R)-configuration of the remaining product. The (R)-configurated 2-hydroxyacid ester
also was obtained by microbial asymmetric hydrolysis of
ethyl 2-acetoxyisocapronate as described by Mori and
Akao (22).

Determination of the enantiomeric composition of chiral aroma constituents in tropical fruits

Tropical fruits, such us passion fruits, mangos or
pineapples, contain many chiral aroma constituents. So
far, their enantiomeric composition is unknown, because
the conventional method, measuring optical rotation, can
not be applied to these components, which can be isolated
from the fruits only in small amounts. The new techniques
of capillary GC analysis of diastereoisomeric derivatives
made it possible to characterize the enantiomeric composition of several chiral trace constituents. These results may be used to gain insight into the biogenesis of
aroma components or to control natural aroma concentrates.

Odd-numbered secondary alcohols (pentanol-2, hepta-
nol-2, nonanol-2) are contained as aroma components in
yellow (Passiflora edulis f. flavicarpa) and purple (Passiflora edulis sims) passion fruits; the corresponding
esters however are typical constituents only of the purple variety (23). The capillary GC investigation of the
enantiomeric composition of these chiral components revealed interesting aspects of their biogenesis.

As shown in Figure 6, pentanol-2 and heptanol-2, isolated by preparative GC from yellow passion fruits, mainly consisted of the (S)-enantiomer; the optical purity of
heptanol-2 was higher than the purity of pentanol-2. In
contrast, heptanol-2 isolated from the purple variety
mainly consisted of the (R)-enantiomer (92 %). Heptanol-2,
obtained by alkaline hydrolysis of the 2-heptylesters in
purple passion fruits, was optically pure and possessed
the (R)-configuration. Thus, the two passion fruit varieties contain different enantiomers of heptanol-2.

The biogenesis of secondary alcohols and their esters
may be explained by a modified pathway of the β-oxidation
of fatty acids (23). The last step in the biosynthesis of
secondary alcohols may be the enzymic reduction of the
corresponding methylketones. In yellow passion fruit this
reduction may be catalyzed by an enzyme, comparable to the
alcohol dehydrogenase in yeast, leading to S-(+)-alkan-
2-ols. The formation of the inverse R-(-)-heptanol-2 in
the purple variety might be catalyzed by another enzyme,
comparable to the dihydroxyacetone reductase isolated
from Mucor javanicus (24) or to the alcohol dehydrogenase
isolated from Thermoanaerobicum brockii (25), which are
known to reduce methylketones to the corresponding R-(-)-
alkan-2-ols. Even though free secondary alcohols of oppo-

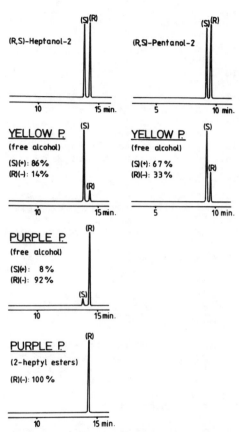

Figure 6. Capillary GC-separation of R-(+)-derivatives of secondary alcohols and their esters, isolated by preparative GC from yellow and purple passion fruits (DB 210, 30 m/0.33 mm i.d., 140 °C, pentanol-2; OV 101, 50 m/0.33 mm i.d., 170 °C, heptanol-2).

site configurations are present in both yellow and purple
passion fruits, the corresponding esters are contained
only in the purple fruits. Because enzyme catalyzed reactions are reversible, it seems propable that in addition
to the specific enzymic hydrolysis of racemic esters of
secondary alcohols to R-(-)-alkan-2-ols only the R-(-)-
alkan-2-ols contained in the purple fruit are converted
to the corresponding esters.

3-Hydroxyacid esters are contained in several sub-
tropical fruits like pineapple (26), passion fruit (27)
and mango (28). 3-Hydroxyacid derivatives are formed as
intermediates during de novo synthesis and β -oxidation
of fatty acids, but the two pathways lead to opposite
enantiomers. S-(+)-3-Hydroxyacyl-CoA-esters result from
stereospecific hydration of Δ 2,3-trans-enoyl-CoA during
β -oxidation; R-(-)-3-hydroxyacid derivatives are formed
by reduction of 3-ketoacyl-S-ACP in the course of fatty
biosynthesis. Both pathways may be operative in the production of chiral 3-hydroxyacids and 3-hydroxyacid esters
in tropical fruits.

The 3-hydroxyacid esters were isolated from the aroma extracts by preparative GC and their enantiomeric composition was determined by capillary GC separation of R-
(+)-MTPA-derivatives. These investigations showed that
the fruits contain different enantiomers of 3-hydroxyacid
esters.

Ethyl 3-hydroxybutanoate, isolated from yellow passion fruit, mainly consisted of the (S)-enantiomer (82 %),
comparable to the product, obtained by yeast reduction of
the corresponding 3-ketoacid ester (see above). In contrast, ethyl 3-hydroxybutanoate in purple passion fruit
and mango mainly consisted of the (R)-enantiomer (69 %
and 78 %).

As shown in Figure 7 ethyl 3-hydroxyhexanoate, isolated from purple passion fruit possessed the (R)-confi-
guration, comparable to the hydroxyacid ester obtained
by the reduction with baker's yeast. In contrary to that
methyl 3-hydroxyhexanoate, which was isolated from aroma
extracts of pineapple, consisted of the (S)-enantiomer
(91 %).

In addition to the 3-hydroxyacid esters, pineapples
also contain the corresponding 3-acetoxyacid esters (29).
Methyl 3-acetoxyhexanoate was isolated by preparative GC,
saponified, methylated with diazomethane and converted to
a R-(+)-MTPA-derivative. As shown in Figure 7, capillary
GC analysis revealed that the 3-acetoxycompound also possessed the (S)-configuration.

It seems that the biogenesis of 3-acetoxyacidesters
in pineapple is an enantio-selective process, comparable
to the formation of esters of secondary alcohols in passion fruits. As the enzymic hydrolysis of ethyl 3-aceto-
xyhexanoate by Candida utilis leads to the (S)-configu-
rated hydroxycompound (see Figure 4), only (S)-3-hydroxy-
acid esters are esterified to the corresponding 3-aceto-
xycompounds in pineapple.

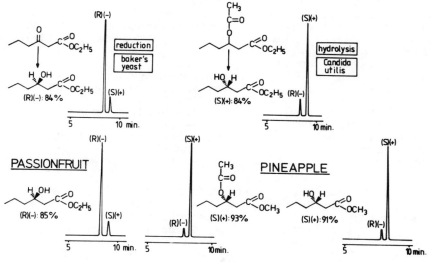

Figure 7. Capillary GC separation of chiral components obtained by microbiological processes and isolated from tropical fruits (DB 210, 30 m/0.33 mm i.d., 185 °C).

In addition to 3-hydroxy- and 3-acetoxyacid esters,
pineapples contain 5-hydroxy- and 5-acetoxyacid esters,
γ -hexalactone and δ -octalactone as characteristic chiral aroma constituents (29). The capillary GC investigation of R-(+)-PEIC derivatives made it possible, to determine the enantiomeric composition of these components
for the first time (Figure 8).

The postulated theory of an enantio-selective biogenesis of acetoxyacid esters in pineapple is further supported by the finding that methyl 5-acetoxyhexanoate
mainly consists of the (S)-enantiomer (79 %), comparable
to methyl 3-hydroxy- and 3-acetoxyhexanoate. γ -Hexalac-
tone, another chiral C_6-component, possesses the (R)-con-
figuration.

It is interesting that in contrast to the chiral
C_8-components, methyl 5-acetoxyhexanoate and δ -octalac-
tone possessed nearly racemic composition. Obviously different biogenetic pathways leading to opposite enantiomers must be operative in the biogenesis of these pineapple constituents. Some possible causes for this stereo-
specificity are enzyme catalyzed hydration of double
bonds, reduction of ketogroups and chain elongation of
the corresponding optically pure precursors as outlined
in Figures 9 and 10.

At the moment, experimental data with labeled precursors are still lacking and the proposed pathways are
to be considered as "working hypothesis". Nevertheless,
the partly unexpected results, which were obtained in
these investigation, show that there must be a competition of pathways and enzymes in the biogenesis of chiral
aroma constituents in tropical fruits.

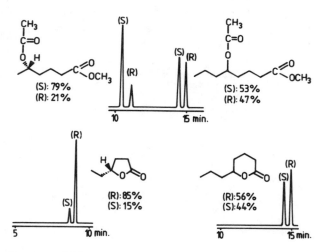

Figure 8. Capillary GC separation of R-(+)-PEIC-deriva-
tives of chiral (main) constituents isolated from pineapple (DB 210, 30 m/0.33 mm i.d., 200 °C).

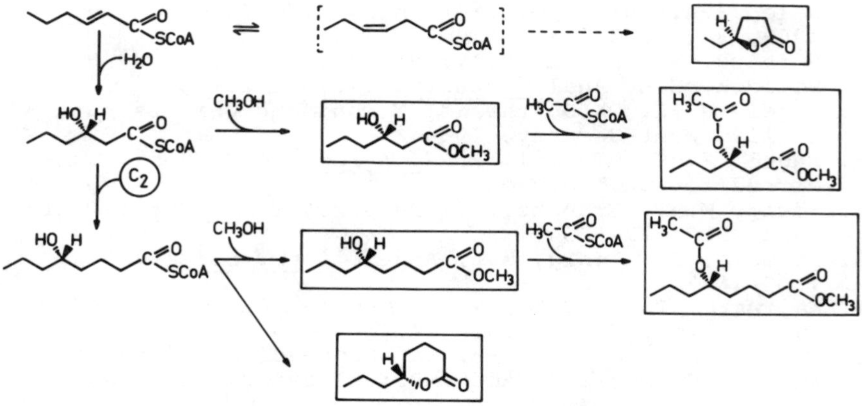

Figure 9. Possible pathway to explain the formation of (S)-3-hydroxyacid esters, (S)-3-acetoxyacid esters, and (S)-5-acetoxyacid asters in pineapple.

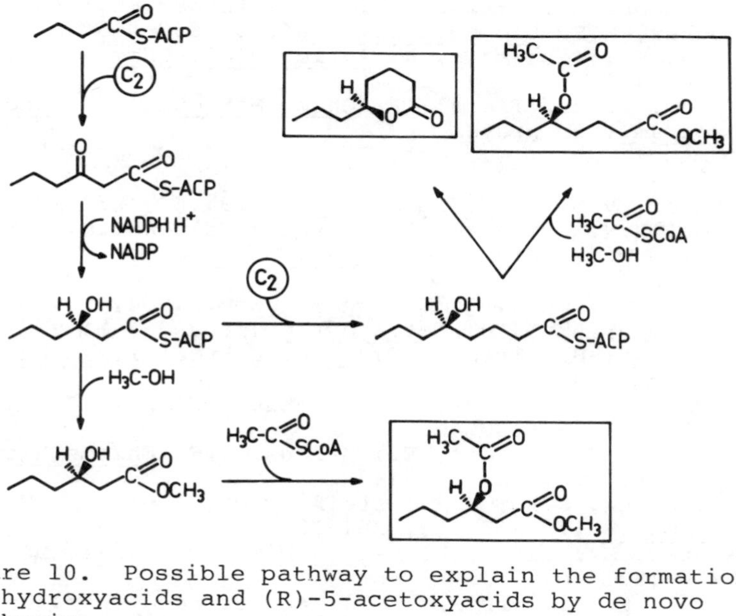

Figure 10. Possible pathway to explain the formation of (R)-hydroxyacids and (R)-5-acetoxyacids by de novo synthesis.

Literature Cited

1. Schurig, V. Angew. Chem. 1984, 96, 733-822
2. Gil-Av, E.; Nurok, D. Advances in Chromatography 1974, 10, 99-172
3. Schurig, V.; Weber, R. Angew. Chem. 1983, 95, 797-798
4. König, W.A. J. HRC & CC 1982, 5, 588-595

5. Dale, J.A.; Dull, D.L.; Mosher, H.S. J. Org. Chem. 1969, 34, 2543-2549
6. Pereira, W.: Bacon, V.A.; Patton, H.; Halpern, B.; Pollock, G.E. Anal. Letters 1970, 3, 23-28
7. Engel, K.-H. Ph.D. thesis, Technische Universität Berlin, West-Germany, 1984
8. Engel, K.-H.; Tressl, R. J. Agric. Food Chem., submitted for publication
9. König, W.A.; Francke, W.; Benecke, I. J. Chromatogr. 1982, 239, 227-231
10. Benecke, I.; König, W.A. Angew. Chem. Suppl. 1982, 1605-1613
11. Kajiwara, T.; Nagata, N.; Hatanaka, A.; Naoshima, Y. Agric. Biol. Chem. 1980, 44, 437-438
12. Tressl, R.; Engel, K.-H. in "Proceedings of the 4. Weurman-Symposium"; Adda, P.; Elsevier: Wageningen, 1984, in press
13. Tressl, R.; Engel, K.-H. in "Analysis of Volatiles"; Schreier, P., Ed.; Walter de Gruyter; Berlin, New York, 1984, pp. 323-342
14. Neuberg, C. Adv. Carbohydrate Res. 1949, 10, 75-117
15. Mac Leod, R.; Prosser, H.; Fikentscher, L.; Lanyi, J.; Mosher, H.S. Biochemistry 1964, 3, 838-846
16. Ohta, H.; Tetsukawa, H. Agric. Biol. Chem. 1980, 44, 863-867
17. Engel, K.-H.; Tressl, R. Chem. Mikrobiol. Technol. Lebensm., in press
18. Lemieux, R.U.; Giguere, J. Canad. J. Chem. 1951, 29, 678-690
19. Frater, V.G. Helv. Chim. Acta 1979, 62, 2825-2829
20. Zhou, B.N.; Gopalan, A.S.; Van Middlesworth, F.; Shieh, W.R.; Sih, C.J. J. Am. Chem. Soc. 1983, 105, 5925-5926
21. Sih, C.J.; Chen, C.-S. Angew. Chem. 1984, 96, 556-565
22. Mori, K.; Akao, H. Tetrahedron 1980, 36, 91-96
23. Engel, K.-H.; Tressl, R. Chem. Mikrobiol. Technol. Lebensm. 1983, 8, 33-39
24. Hochuli, E.; Taylor, K.E., Dutler, H. Eur. J. Biochem. 1977, 75, 433-439
25. Lamed, R.J.; Keinan, E.; Zeikus, J.G. Enzyme Microb. Technol. 1981, 3, 144-148
26. Creveling, R.K.; Silverstein, R.M.; Jennings, W.G. J. Food Sci. 1968, 33, 284-287
27. Winter, M.; Klöti, R. Helv. Chim. Acta 1972, 55, 1916-1921
28. Engel, K.-H.; Tressl, R. J. Agric. Food Chem. 1983, 31, 796-801
29. Näf-Müller, R.; Willhalm, B. Helv. Chim. Acta 1971, 54, 1880-1890

RECEIVED September 16, 1985

A New Analytical Method for Volatile Aldehydes

Tateki Hayashi[1], Clayton A. Reece, and Takayuki Shibamoto

Department of Environmental Toxicology, University of California, Davis, CA 95616

Trace quantities of formaldehyde and methyl glyoxal in
aqueous and food samples were determined by a newly developed method. Formaldehyde and methyl glyoxal were reacted
with cysteamine in aqueous medium or food sample to give
thiazolidine and 2-acetylthiazolidine, respectively, at pH
6 and 8. Thiazolidine derivatives formed from formaldehyde and methyl glyoxal were extracted with dichloromethane or chloroform and subsequently analyzed by a gas chromatograph equipped with a fused silica capillary column
and a thermionic detector. Seventeen commercial food
items were analyzed for formaldehyde and methyl glyoxal.
The quantities of formaldehyde and methyl glyoxal varied
from 0 to 17 ppm and from 0 to 620 ppm, respectively.

Certain volatile aldehydes such as formaldehyde and methyl glyoxal
have always presented some difficulties in the determination of
their levels in foods and beverages. Formaldehyde is difficult to
extract from an aqueous solution with an organic solvent because it
is very water soluble or exists as a polymer in an aqueous media.
Methyl glyoxal is also hard to recover from food samples because it
exists as a copolymer with some amines such as amino acids and
proteins.

Formaldehyde is widely used in many manufacturing processes and
its production in the U.S. reached 5.6 billion pounds in 1980 (1).
Exposure to formaldehyde has caused dermatitis and pulmonary
irritation in workers, and recent evidence based on animal studies
has implicated formaldehyde as a potential carcinogen (2).

Methyl glyoxal has been found in many foods, such as bread (3),
boiled potatoes (4), roast turkey (5), and tobacco smoke (6). It is
a well known fact that sugar caramelization produces numerous
carbonyls including formaldehyde and methyl glyoxal (7). Among

[1]Current address: Nagoya University, Department of Food Science and Technology,
Nagoya, Japan 464

those products, methyl glyoxal is one of the most highly reactive compounds and readily undergoes secondary reactions to form some heterocyclic compounds (8).

A browning model system consisting of cysteamine and D-glucose produced numerous thiazolidine derivatives (9). It was proposed that D-glucose decomposed into short-chain carbonyls such as formaldehyde, acetaldehyde, glyoxal, and methyl glyoxal and subsequently reacted with cysteamine to give the corresponding thiazolidine derivatives. This suggests that cysteamine reacts readily with carbonyl compounds to yield thiazolidine derivatives in an aqueous solution. Thiazoles, dehydrated products of thiazolidine, have been found in various foods (10), but thiazolidines have never been isolated from any food samples. The formation of thiazolidine from formaldehyde and cysteamine was reported 50 years ago (11). Some alkylthiazolidines were prepared from cysteamine and aldehydes to synthesize 2-alkyl-N-nitrosothiazolidines for the mutagenicity study of nitrosamines (12).

Since direct analyses for formaldehyde and methyl glyoxal are difficult with gas chromatography (GC) or any other methods, we attempted to determine levels of formaldehyde and methyl glyoxal in various food samples using their derivatives thiazolidine and 2-acetylthiazolidine, respectively. The proposed mechanism of thiazolidine formation from cysteamine and corresponding aldehydes is shown in Figure 1.

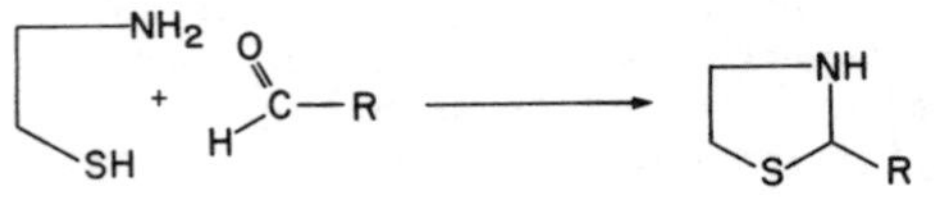

Figure 1. Proposed formation mechanism of thiazolidines.

Literature Review

Formaldehyde. Formaldehyde is one of the most common aldehydes in foods and the environment. Formaldehyde is also an important air pollutant in a variety of industrial and domestic atmospheres (13).

Formaldehyde is present in various forms in an aqueous medium. The simplest form is free (monomeric) formaldehyde which is known to be toxic. Other forms of formaldehyde, such as a polymer or a copolymer with other compounds, may produce toxic free formaldehyde under certain conditions. These forms of formaldehyde are categorized as follows:

1. formaldehyde adsorbed onto particles.
2. para (polymeric) formaldehyde.
3. formaldehyde that has combined with other compounds to produce identifiable substances such as hexamine, which may be in solution or absorbed onto particles.
4. formaldehyde that has combined with, for example, proteins which may then remain in solution or be adsorbed onto other particles.

In spite of the increasing interest in monitoring the levels of formaldehyde in foods and the environment. there is no satisfactory, simple method available for the determination of trace quantities of formaldehyde. Many laboratories have attempted to develop a method for determining trace quantities of formaldehyde in various samples. The conventional methods used most widely are shown in Table I.

Table I. Conventional methods used for formaldehyde analysis

	Method	Determinand	Range of concentration (mg/1)	Basis
A.	Manual	Free	0-2	Spectrophotometry
B.	Manual	Total	0-10	Distillation and spectrophotometry
C.	Automated	Free	0-25	Spectrophotometry
D.	Automated	Total	0-25	Direct hydrolysis and spectrophotometry
			0-5	Distillation and spectrophotometry
E.	Automated	Methanol	0-20	Oxidation and spectrophotometry
F.	Manual	Methanol	0-50	Gas chromatogrphy with FID

Free formaldehyde is reacted with acetylacetone in the presence of an excess of an ammonium salt to form the yellow fluorescent compound, 3,5-diacetyl-1,4-dihydrolutidine and subsequently determined spectrophotometrically in methods A-E (14). In these methods, the test sample must be colorless and free from other carbonyl compounds. Some other derivatives have been used to analyze formaldehyde. For example, formaldehyde was reacted with sodium 4,5-dihydroxy-2,7-naphthalene disulfonate in sulfuric acid solution to yield a purple color (580 nm) and then subjected to colorimetric analysis. A purple-colored pararosaniline derivative was used to analyze formaldehyde in air (15). Air sample was passed through an aqueous solution which contained 0.4% of 3-methyl-2-benzothiazolone hydrazone hydrochloride and then a dye produced was determined at 635 or 670 nm (16). Molecular sieve (1.6 mm pettet) was used to trap formaldehyde in air samples. The formaldehyde trapped on the molecular sieve was rinsed off with deionized water and subsequently determined colorimetrically as a pararosaniline derivative. The recovery of ppb levels was reported using this method (17). A major drawback of these colorimetric methods is that many other compounds can interfere with the analysis. If a test solution is contaminated with a compound having an absorption around 570-580 nm, significant interference occurs.

Formaldehyde in air was sampled with silica gel coated with 2,4-dinitrophenylhydrazine (2,4-DNPH) and the resulting hydrazone was extracted with acetonitrile and determined by reverse-phase HPLC with UV detection at 340 nm. This method was validated over the range of 2.5-93.3 μg formaldehyde (18). An air sample was bubbled into an aqueous solution of 2,4-DNPH and the hydrazones that formed were determined by high performance liquid chromatography (HPLC). Fourteen aldehydes including formaldehyde were analyzed in an air sample using this method (19). Formaldehyde in fresh shrimp was analyzed by HPLC as 2,4-DNPH derivative. Characteristics of this method included an estimated detection limit of 0.05 mg of formaldehyde/kg of shrimp, an average recovery of 72.3% at the 10 mg/kg level, and a total analysis time of 2 h. (20). Complete GC separation of the 2,4-DNPH derivatives of ten aliphatic aldehydes, eight aliphatic ketones and four aromatic aldehydes was obtained with a 20 m x 0.25 mm i.d. glass capillary column coated with SF-96, with the exception of the derivatives of n-valeraldehyde and isobutyl methyl ketone, whose peaks overlapped, and the o- and m-tolualdehyde derivatives, which were poorly separated (21). High resolution glass capillary GC separate syn-anti isomers of 2,4-DNPH derivative of volatile aldehydes (22). Sample collection from a automobile exhaust and derivatization were performed directly in a midget impinger containing an acetonitrile solution of 2,4-DNPH and catalyst. This method allowed direct injection of an aliquot of the sample into HPLC. The detection limit for formaldehyde was 20 ppb with an analysis time as short as 10 min (23). Ambient air samples were collected on molecular sieve 13X absorbents at 2-h intervals, and the trapped formaldehyde was then determined by the mass fragmentograms of m/z 29 and m/z 30. The detection limit of this method was sufficient to quantify the low ppb levels of ambient formaldehyde in rural air (24). Formaldehyde in air reacted with N-benzylethanolamine-coated Chromosorb 102 sorbent to produce a derivative of formaldehyde, 3-benzyloxazolidine. The oxazolidine recovered from the sorbent was determined using a GC equipped with a 25m fused silica capillary column coated with Carbowax 20M. The detectable range of this method was from 0.55 to 4.71 mg/m^3 (25).

Methyl glyoxal. Methyl glyoxal (pyruvaldehyde, 2-ketopropionic aldehyde, or 2-oxopropanal), is a hygroscopic, yellow, mobile liquid with a pungent, stinging odor, was first obtained by warming isonitrosoacetone with dilute sulfuric acid in 1887 (26). As mentioned above, methyl glyoxal is one of the sugar caramelization or decomposition products and is often found in heat-treated food. Methyl glyoxal and a number of other carbonyl compounds were isolated from sucrose melted at 150°C, the temperature of baking bread crusts (27). One cup of instant coffee (1g/100 ml of water) and one cup of coffee prepared from ground coffee beans (8g/100 mL of water) contained 100-150 ug and 470-730 ug of methyl glyoxal, respectively (28). Methyl glyoxal has been used in minute traces in imitation coffee, maple, honey, caramel and rum flavor (29).

Direct analysis of methyl glyoxal by gas chromatography is possible; however, it may be present as a copolymer with other compounds in foods so aht recovery efficiency of methyl glyoxal by organic solvent extraction has not been established. Methyl glyoxal and other dicarbonyl compounds in cigarette smoke were analyzed with GC or HPLC as quinoxaline derivatives after reacting with o-phenylenediamine. The levels of methyl glyoxal in the smoke from commercial cigarettes were 33-70 ug/cigarett and 19-40 ug/cigarette for non-filter cigarettes and filter cigarettes, respectively (6). This method was also used to determine levels of methyl glyoxal in coffee samples (28). There is no simple method available for methyl glyoxal analysis at the present time.

Experimental

Materials. Cysteamine hydrochloride, formaldehyde (37% in water), methyl glyoxal (40% in water), and N-methylacetamide were purchased from the Aldrich Chemical Co., Milwaukee, WI. The extraction solvents (dichloromethane and chloroform) were obtained commercially and used without further treatment. Standard fatty aldehydes were obtained from reliable commercial sources.

Instrumental Analysis. A Hewlett-packard Model 5880 A GC, equipped with thermionic specific detector and a 50 m x 0.23 mm i.d. fused silica capillary column coated with Carbowax 20M, was used for quantitative analysis of thiazolidine and 2-acetylthiazolidine derived from formaldehyde and methyl glyoxal, respectively. GC peak areas were calculated with a HP 5880 A series GC integrator. The oven temperature was programmed from 70 to 180°C at 2° C/min. A Finnigan Model 3200 combination GC/MS equipped with an INCOS MS data system was used for mass spectral identification of thiazolidine derivatives.

GC analysis of volatile aldehyde standards. A mixture of formaldehyde, acetaldehyde, propionaldehyde, isobutyl aldehyde, isovaleraldehyde, methyl glyoxal, and furfural (0.1 mg each) were added to 20 ml of cysteamine solution (6g/1 liter of deionized water). The pH of the solution was adjusted to 8 with 6 N NaOH solution. The reaction proceeded promptly to formm thiazolidine derivatives. The reaction mixture was then extracted with 2 ml of dichloromethane, and an aliquot of the extract was injected in the GC. A gas chromatoram of the extract is shown in Figure 2.

Components were identified with GC/MS. A thiazolidine derivative is easily identified using single ion. monitoring with m/z 88 (thiazolidine ring - H). Table II shows mass spectra and GC retention data of thiazolidine derivatives. The gas chromatogram of the extract from the reaction mixture indicated that methyl glyoxal produced three products, 2-acetylthiazoline (peak #7), 2-acetylthiazolidine (peak #8), and 2-formyl-2-methylthiazolidine (peak #9).

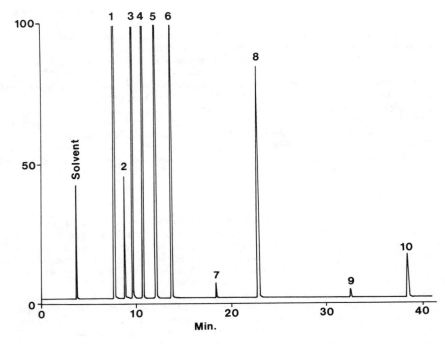

Figure 2.   Gas chromatogram of standard thiazolidines. Peak 1 = 2-
methylthiazolidine, 2 = thiazolidine, 3 = 2-ethylthiazolidine, 4 =
2-isopropylthiazolidine, 5 = N-methylacetamide (internal standard),
6 = 2-isobutylthiazolidine, 7 = 2-acetylthiazoline, 8 = 2-acetyl-
thiazolidine, 9 = 2-formyl-2-methylthiazolidine, 10 = 2-(2-furyl)-
thiazolidine.

Table II. The products of aldehydes and cysteamine and their
           spectral data

Peak # in Figure 3	Products	MS data
1	2-Methylthiazolidine	M^+ = 103 (48, 88 (76), 56 (100)
2	Thiazolidine	M^+ = 89 (100), 88 (43), 59 (38)
3	2-Ethylthiazolidine	M^+ = 117 (21), 88 (100), 70 (21)
4	2-Isopropylthiazolidine	M^+ = 131 (5), 88 (100)
5	N-Methylacetamide	
6	2-Isobutylthiazolidine	M^+ = 145 (8), 88 (100), 56 (37)
7	2-Acetylthiazoline	M^+ = 129 (100), 101 (11), 60 (76)
8	2-Acetylthiazolidine	M^+ = 131 (4), 88 (100), 61 (27)
9	2-Formyl-2-methyl-thiazolidine	M^+ = 13 (4), 101 (32), 60 (100)
10	2-(2-furyl)thiazolidine	(tentative)

It would be ideal if 2-acetylthiazolidine was formed
exclusively from methyl glyoxal. The optimum reaction condition for
2-acetylthiazolidine, therefore, was determined by the following
experiments:

Reaction temperature. The same molar ratio of methyl glyoxal and
cysteamine was reacted at 0, 25, and 100°C. The results are shown
in Table III. The reaction at room temperature (20°C) gave the best
results for 2-acetylthiazolidine formation.

Table III. Relative ratio of products from the reaction of methyl
            glyoxal and cysteamine at various temperature

Temperature (° C)	Products ratio (%)		
	2-Acetyl-thiazoline	2-Acetyl-thiazolidine	2-Formyl-2-methyl thiazolidine
0	81.0	19.0	0
25	8.5	89.0	2.5
100	2.0	40.0	58.0

The effect of molar ratio of methyl glyoxal and cysteamine. This
was examined at room temperature and the results are shown in Table
IV. When the molar ratio of cysteamine and methyl glyoxal was 1000
at pH 8, 2-acetylthiazolidine was produced exclusively. On the
other hand, formation of 2-acetylthiazolidine remained constant pH
6. Therefore, cysteamine was reacted with samples of interest at
25°C and in a quantity to exceed 1000 fold the estimated amount of
methyl flyoxal in the following experiments, when experiment was
cunducted at pH 8.

Table IV. Effect of reactants ratio on 2-acetylthiazolidine
           formation from cysteamine and methylglyoxal

Cysteamine Methylglyoxal (Molar ratio)	Yield of products (%)					
	2-Acetyl-thiazoline		2-Acetyl-thiazolidine		2-Formyl-2-methyl-thiazolidine	
	pH 6	pH 8	pH 6	pH 8	pH 6	pH 8
1	2.5	1.6	96.8	98.8	0.7	0.2
10	2.0	1.2	98.0	47.5	0.0	51.3
100	2.2	0.0	97.8	63.0	0.0	37.0
1000	0.0	0.0	98.2	100.0	1.8	0.0

<u>Effect of pH</u>. The effect of pH on thiazolidine and 2-
acetylthiazolidine was examined using solutions (250 ml) containing
cysteamine (1.5 g)/formaldehyde (0.198 mg) and cysteamine (0.75
g)/methyl glyoxal (o.57 mg), respectively. The optimum recovery
efficiency of thiazolidine and 2-acetylthiazolidine was obtained
around pH 8. The test solutions were, therefore, adjusted to pH 8
roiutinely. The recovery of thiazolidine reached its maximum at pH
8 and leveled off (Figure 3). Assuming that the reaction proceeds
at 100% efficiency, 0.1975 mg of formaldehyde (MW = 30) should
produce 0.5859 mg of thiazolidine (MW = 89). The maximum
formaldehyde recovery calculated as thiazolidine from the
calibration curve was 118% at pH 8. The reason(s) for this excess
recovery is not yet understood. These data are averages of at least
three replications. The correction can, therefore, be made easily
for an actual analysis. In another series of experiments, methyl
glyoxal was recovered significantly less from a solution of pH 6
than from a solution of pH 8.

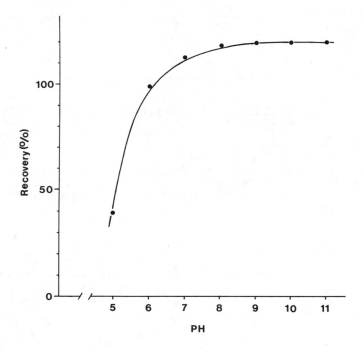

Figure 3. Effect of pH on thiazolidine recovery.

<u>Preparation of calibration curve for formaldehyde analysis</u>. The
calibration curve for thiazolidine (formaldehyde derivative) was
prepared with N-methylacetamide as an internal standard.

N-Methylacetamide (0.5 mg) was added to each standard chloroform
solution of thiazolidine (0.125-1.00 mg). Gas chromatographic peak
area ratio of thiazolidine and the standard were plotted against
quantity of thiazolidine. A typical calibration curve prepared for
thiazolidine determination is shown in Figure 4.

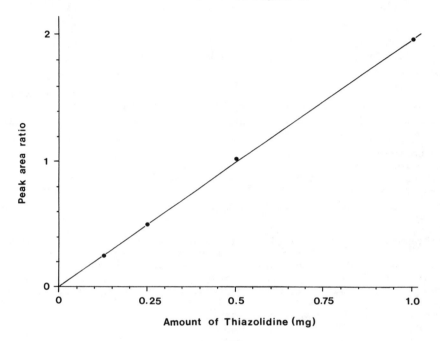

Figure 4. A gas chromatographic calibration curve for thiazolidine.

Preparation of calibration curve for methyl glyoxal analysis. The
calibration curve for methyl glyoxas was prepared using N-
methylacetamide as an internal standard. N-Methylacetamide was
added to each standard reaction mixture of methyl glyoxal (1.0-7.5
mg) and 0.75 g of cysteamine in 70 ml of dichloromethane at pH 6.
Gas chromatographic peak ratio of 2-acetylthiazolidine and the
standard were plotted against quantity of methyl glyoxal in the
original solution.

Solvent Choice. When an aqueous solution of cysteamine (1g/250 ml,
pH 8) was extracted with dichloromethane using a liquid-liquid
continuous extractor for three hours, a certain amount of
thiazolidine was isolated and identified by GC/MS. The presence of
thiazolidine in a dichloromethane extract was observed even though
the cysteamine solution was washed with ethyl acetate prior to
use. The thiazolidine also was isolated when double distilled water
was used instead of deionized water to prepare cysteamine
solutions. Thiazolidine quantity decreased significantly when
chloroform or ethyl acetate was used as an extraction solvent.

These results indicated that dichloromethane contains certain quantities of formaldehyde as a contaminant. Formaldehyde contamination in solvents was confirmed by the following experiment. Various concentrations of cysteamine solution were extracted using different quantities of dichloromethane. The conditions and results of this experiment are show in Table V.

Table V. Effect of solvent volume and cysteamine quantity on thiazolidine recovery from solvents

Solvent	Volume (ml)	Amount of cysteamine added (g)	Thiazolidine recovered (mg)
Dichloromethane	300	0.60	14.0
Dichloromethane	100	0.70	5.8
Dichloromethane	100	0.25	4.3
Chloroform	100	2.1	0.033
Chloroform	100	0.75	0.033

When the quantity of dichloromethane was reduced from 300 ml to 100 ml, the amount of thiazolidine recovered decreased from 14 mg to 4.3-5.8 mg. When chloroform was used as a solvent, the amount of thiazolidine recovered remained constant (0.0033 mg).

Reagent grades of dichloromethane and chloroform obtained from various commercial sources were analyzed for formaldehyde in order to choose the optimum solvent for further experiments. The results were shown in Table VI.

Table VI. Quantities of formaldehyde in commercial dichloromethane and chloroform

Solvent	Source	Quantity of formaldehyde (ppm)
Dichloromethane	A_1	14.0
Dichloromethane	A_2	9.1
Dichloromethane	B	2.3
Dichloromethane	C	12.0
Dichloromethane	D	9.4
Chloroform	A_1	0.064
Chloroform	A_2	0.15
Chloroform	A_3	0.11
Chloroform	B	0.77

[1,2,3]Different batch

All dichloromethane examined showed 2-14 ppm of formaldehyde contamination. Several clean up methods were applied to remove formaldehyde such as washing with sodium bisulfite, treatment with active charcoal of Porapak Q porous polymer without success. Trace levels of formaldehyde in solvents may be impossible to remove. Therefore, chloroform was used as the solvent for formaldehyde analysis in further experiments. The amount of contaminant obtained from a blank solvent was always subtracted from the values of actual results. Dichloromethane was, however, used for methyl glyoxal analysis. The extraction efficiency of chloroform and dichloromethane was almost identical. Dichloromethane was easier to use for a liquid-liquid continuous extraction than chloroform because of its lower boiling point.

Sample preparations

1) <u>D-Glucose</u> the system consisting of methyl glyoxal and D-glucose was examined to determine interference of D-glucose in thiazolidine and 2-acetylthiazolidine formation. The solutions containing formaldehyde, methyl glyoxal and D-glucose were treated with cysteamine. The reaction conditions are shown in Table VII along with the recovery efficiencies.

2) <u>Soy sauce</u> cysteamine (0.75g) was added to 20 ml of soy sauce.

3) <u>Soy bean paste</u> soy bean paste (20 g) was dissolved in 200 ml of deionized water and 0.75 g of cysteamine was added.

4) <u>Brewed coffee</u> regular or decaffeinated coffee (80 g) was added to 1 liter of boiling water. After 10 min., the coffee was filtered and 250 ml of filtrate was reacted with 0.75 g of cysteamine.

5) <u>Instant coffee, cocoa, instant tea, maple syrup, and nonfat dry milk</u> instant coffee (3 g), cocoa (5 g), maple syrup (10 g), and nonfat dry milk (8 g) were dissolved in 250 ml each of hot deionized water and each solution was reacted with 0.75 g of cysteamine.

6) <u>Coke, wine, beer, orange juice, tomato juice, root beer, and apple juice</u> Cysteamine (0.75 g) was added directly to 250 ml of each of these samples.

All of the above solutions were stirred for 30 min with a magnetic stirrer. The pH of the solutions was adjusted to 6 or 8 with 6N NaOH immediately after addition of cysteamine. The reaction mixtures were extracted with 70 ml of dichloromethane or chloroform for 6 h using a liquid-liquid continuous extractor. The extracts were dried over anhydrous sodium sulfate for 12 h. After the removal of sodium sulfate, 0.5 mg of N-methylacetamide was added to each solution as an internal standard. The extracts were quantitatively analyzed for thiazolidine and 2-acetylthaizolidine by GC.

Results and Discussion

Preliminary experiments. The results of preliminary experiments showed that volatile aldehydes reacted with cysteamine readily to give the corresponding thiazolidine derivative. Thiazolidine derivatives are much more stable than underivatized formaldehyde or methyl glyoxal. Solvent contaminants did not interfere with the GC analysis of thiazolidine. Moreover, because thiazolidine contains nitrogen, a highly sensitive and selective thermionic detector is applicable for analysis. When a food sample is treated with cysteamine, some food constituents such as carbohydrate may interfere with thiazolidine or 2-acetylthiazolidine formation. D-Glucose was chosen to represent a possible interference caused by food constituents. Recovery of formaldehyde and methyl glyoxal was reduced about 10% and 5% in the presence of 10% D-glucose, respectively (Table VII). In the 10% D-glucose solution, the quantity of D-glucose was over 10,000 times that of formaldehyde of methyl glyoxal. Food constituents, such as D-glucose, apparently do not interfere significantly with thiazolidine or 2-acetylthiazolidine formation.

Table VII. Recovery efficiency of thiazolidine and 2-acetylthiaz-
            olidine in the presence of D-glucose.

Aldehyde	Amount used (mg)	Concentration of glucose (%)	Product	Yield of product (%)
Methylglyoxal	0.57	0	2-Acetyl-thiazolidine	100
Methylglyoxal	0.57	5	2-Acetyl-thiazolidine	96
Methylglyoxal	0.57	10	2-Acetyl-thiazolidine	95
Formaldehyde	0.2	0	Thiazolidine	100
Formaldehyde	0.2	10	Thiazolidine	90

Analysis of formaldehyde and methyl glyoxal in food samples. The chloroform extract of cysteamine-treated decaffeinated coffee showed obvious existence of aldehydes (Figure 5). In contrast to the gas chromatogram of cysteamine-untreated decaffeinated coffee (Figure 6), that of the treated sample showed new peaks; 1 (2-methylthiazolidine, 2 (thiazolidine), 3 (2-acetylthiazolidine), and 4 (furfurylthiazolidine). The peak of the internal standard (S) did not interfere with any thiazolidine derivatives.

The number of peaks appearing in coffee extracts seemed to be too small but this was due to high dilution of the samples. The number of peaks appearing in the coffee extracts seemed to be too small, compared to their number in a previous study using concentrated samples (30, 31) instead of the solvent-diluted samples of the present study. The relatively high dilution of these samples would explain the smaller number of peaks. Among the coffee

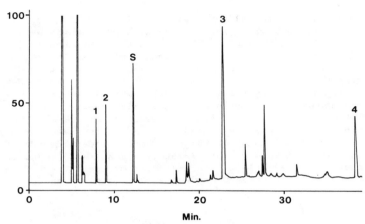

Figure 5. A gas chromatogram of the chloroform extract of cysteamine-treated decaffeinated coffee. Peak 1 = 2-methylthiazolidine, 2 = thiazolidine, S = internal standard, 3 = 2-acetylthiazolidine, 4 = 2-furfurylthiazolidine.

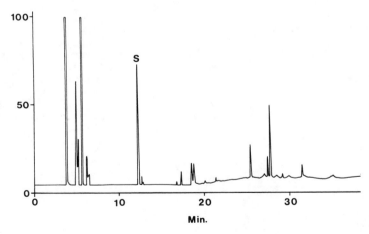

Figure 6. A gas chromatogram of the chloroform extract of cysteamine-untreated decaffeinated coffee.

volatiles, nitrogen-containing compound such as pyrazines and pyrroles, may be, detected by the thermionic detector and appear as peaks on the chromatogram.

The results of formaldehyde and methyl glyoxal analysis in commercial foods are shown in Table VIII. Formaldehyde was identified in the levels of 3.7-17 ppm in coffee obtained from various commercial sources. It was found at higher levels in instant coffees than in brewed coffee. This suggests that formaldehyde may escape from coffee during brewing. Formaldehyde has been reported in coffee volatiles by several researchers (31). There are however, no reports on quantitative analysis of formaldehyde in coffee prior to the present study.

Methyl glyoxal has never been reported in soy sauce or soy bean paste prior to this study. Certain aldehydes (acetaldehyde, n-propanal, 2-methylpropanal, and 3-methylbutanal) were found in soy sauce previously (32). A gas chromatogram of the extract from cysteamine-treated soy sauce and untreated soy sauce are shown in Figures 7 and 8.

Table VIII. Formaldehyde and methylglyoxal contents in foods

| | Content (ppm) | | |
| | Formaldehyde | Methylglyoxal | |
Food	pH 6	pH 6	pH 8
Group I			
Brewed coffee	4.9	25.0	540
Decaffeinated brewed coffee	3.7	47.0	620
Instant coffee A	10.0	23.0	430
Instant coffee B	17.0	ND	430
Cocoa	3.0	1.2	0
Instant tea	3.0	2.4	0
Nonfat dry milk	1.8	1.4	0
Soy Sauce A	1.2	7.6	66
Soy Sauce B	0.88	3.0	87
Soy bean paste (miso)	3.5	0.7	5.0
Group II			
Coke A	0.35	0.23	0.4
Coke B	ND	0.24	ND
Root beer	0.44	0.76	2.3
Beer	0.08	0.084	0.57
Wine (white)	0.096	0.11	0.28
Apple juice	0.12	0.26	0.31
Orange juice	0.15	0.04	0.39
Tomato juice	0	0.064	0.11
Maple syrup	1.5	2.5	12.0

ND:Data not available.

The additional peaks in Figure 7 represent thiazolidine derivatives. The content of methyl glyoxal in soy bean paste was
considerably lower than that of soy sauce even though they were
prepared by similar procedures. This may be due to differences in
fermentation time.

Coffee contained the largest quantity of methyl glyoxal among
the food samples tested. The values of instant coffee were less

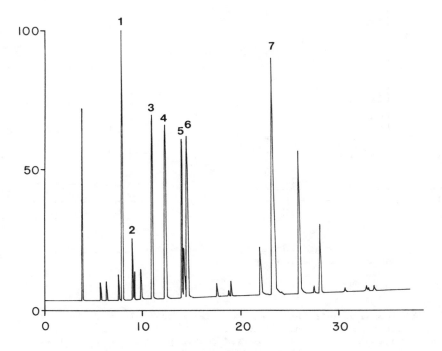

Figure 7. A gas chromatogram of the chloroform extract of
cysteamine-treated soy sauce. Peak 1 = 2-methylthiazolidine, 2 =
thiazolidine, 3 = 2-isopropylthiazolidine, 4 = internal standard
(N-methylacetamide, 5 = isobutylthiazolidine, 6 = unknown thiazolidine derivate, 7 = 2-acetylthiazolidine.

than that of brewed coffee. This is consistent with the results
reported previously (28). In the present study, one cup of instant
coffee (1 g/100 ml) and brewed coffee (8 g/100 ml) contained 238 ug
and 900-1030 ug of methyl glyoxal, respectively, when they were
treated with cysteamine at pH 8. On the other hand, methyl glyoxal
recovery was reduced significantly when the coffee was treated with
cysteamine at pH 6. The same phenomenon was observed in the case of
soy sauce. It is not clear why more methyl glyoxal was recovered at
pH 8. Methyl glyoxal may exist as a polymer or may form a complex,

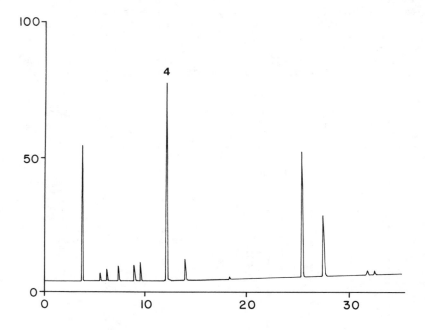

Figure 8. A gas chromatogram of the chloroform extract of
cysteamine-untreated soy sauce.

combining with an amino group of a large molecule such as a protein
in a lower pH solution. At a higher pH such as 8, a certain amount
of methyl glyoxal may be released from a polymer or a complex.

It was expected that foods containing a caramelized sugar such
as coke and maple syrup would have more methyl glyoxal (33).
However, the amount of methyl glyoxal detected in these products was
less than that of coffee or soy sauce.

The samples examined in this study can be classified into two
groups. One is foods consumed without additional water (Group I)
and the other is foods consumed with addition of a certain amount of
water (Group II). It is important to know the amount of
formaldehyde and methyl glyoxal intake when a food is consumed.
Table IX shows the calculated values of methyl glyoxal intake for
each food item when it is consumed. Even after dilution, coffee
provided the most methyl glyoxal intake. It is difficult to
estimate formaldehyde or methyl glyoxal intake from soy sauce
because its use varies widely.

Since formaldehyde and methyl glyoxal have been proven to be
genotoxic and are found in various foods and beverages in somewhat
significant quantities, it is important to study further their risk
to human health.

Table IX.   Calculated values of formaldehyde (FA) and methylglyoxal
            (MG) intake for each food when it is consumed

| Food | Amount consumed | Amount of intake (µg) | | |
| | | FA | MG | |
		pH 6	pH 6	pH 8
Group I				
Brewed coffee	3g/180ml	14.8	75.6	1620
Decaffeinated coffee	3g/180ml	11.2	140.4	1854
Instant coffee A	1g/180ml	10.1	22.7	428
Instant coffee B	1g/180ml	17.1	ND	428
Cocoa	4g/180ml	12.1	4.9	0
Instant tea	0.3g/180ml	0.9	0.7	0
Nonfat dry milk	22.7g/240ml	30.6	31.2	0
Group II				
Coke A	354ml/can	123.9	81.4	141.6
Coke B	354ml/can	ND	85.0	ND
Root beer	354ml/can	155.8	269.0	814.2
Beer	355ml/can	28.4	29.7	201.8
Wine (white)	100ml/glass	9.6	11.0	28.0
Apple juice	300ml/glass	36.0	78.0	31.9
Orange juice	354ml/can	53.1	14.2	138.1
Tomato juice	177ml/can	0	11.3	19.5

ND:Data not available.

Acknowledgments

The financial support of this study by the University of California
Cancer Research Coordinating Committee (#3-504017-19900) is
gratefully acknowledged.

Literature Cited

1.   Chem. Eng. News 1981, 59, 35.
2.   Karns, W. D. "Long-Tern Inhalation Toxicity and Carcinogen-
     icity Studies of formaldehyde in Rats and Mice." The Third
     Chemical Industry Institute of Toxicology Conference on
     Toxicology, Raleigh, NC, 1980.
3.   Wiseblatt, L.; Kohn, F. E. Cereal Chem. 1960, 37, 55-56.
4.   Kajita, T.; Senda, M. Nippon Nogei Kagaku Kaishi. 1972, 46,
     137-145.
5.   Hrdlicka, J.; Kuca, J.; Poult. Sci. 1965, 44, 27-31.
6.   Moree-Testa, P.; Saint-Jalm, T. J. Chromatogr. 1981, 217,
     197-208.
7.   Hodge, J. E. Symp. Foods: Chem. Physiol. Flavors, Proc.
     1967, p. 472.
8.   Ohnishi, S.; Shibamoto, T. J. Agric. Food Chem. 1984, 32,
     987-992.

9.   Sakaguchi, M.; Shibamoto, T. J. Agric. Food Chem. 1978, 26, 1179-1183.
10.   Maga, J. A. Crit. Rev. Food Sci. Nutri. 1975, 153-176.
11.   Ratner, S.; Clarke, H. T. J. Am. Chem. Soc. 1937, 59, 200-207.
12.   Umano, K.; Shibamoto, T.; Fernando, S. Y.; Wei, C.-I. Food Chem. Toxicol. 1984, 22, 253-259.
13.   Matthews, T. G.; Howell, T. C. Anal. Chem. 1982, 54, 1495-1498.
14.   Nash, T. Biochem. J. 1953, 55, 416.
15.   Miksch, R. R.; Anthon, D. W.; Hollowell, C. D.; Revzan, K.; Glanville, J. Anal. Chem. 1981, 53, 2118-2123.
16.   Sawicki, E.; Hauser, T. R.; Stanley, T. W.; Elbert, W. Anal. Chem. 1961, 33, 93-96.
17.   Georghiou, P. E.; Harlick, L.; Winsor, L.; Snow, D. Anal. Chem. 1983, 55, 567-570.
18.   Beasley, R. K.; Hoffmann, C. E.; Rueppel, M. L.; Worley, J. W. Anal. Chem. 1980, 52, 1110-1114.
19.   Kuwata, K; Uebori, M.; Yamasaki, Y. J. Chromatogr. Sci. 1979, 17, 264-268.
20.   Radford, T.; Dalsis, D. E. J. Agric. Food Chem. 1982, 30, 600-602.
21.   Hoshika, Y; Takata, Y. J. Chromatogr. 1976, 120, 379-389.
22.   Uralets, V. P.; Rijks, J. A.; Leclercq, P. A. J. Chromatogr. 1980, 194, 135-144.
23.   Lipari, F.; Swarin, S. J. J. Chromatogr. 1982, 247, 297-306.
24.   Yokouch, T.; Fuji, T.; Ambe, Y.; Fuwa, K. J. Chromatogr. 1979, 180, 133-138.
25.   Kennedy, E. R.; Hill, R. H. Jr. Anal. Chem. 1982, 54, 1739-1742.
26.   Pechmann, V. Chem. Ber. 1887, 20, 3213.
27.   Hodge, J. E. J. Agric. Food Chem. 1953, 1, 928-943.
28.   Kasai, H.; Kumeno, K.; Yamaizumi, Z.; Nishimura, S.; Nagao, M.; Fujita, Y.; Sugimura, T.; Nukaya, H.; Kosuge, T. Gann. 1982, 73, 681-683.
29.   Arctander, S. "Perfume and Flavor Chemicals"; Steffen Archtander: Elizabeth, N. J., 1969; p. 2786.
30.   Shibamoto, T.; Harada, K.; Mihara, S.; Nishimura, O.; Yamaguchi, K.; Aitoku, A.; Fukada, T. Application of HPLC for evaluation of coffee flavor quality. In the quality of food beverages. G. Charalambous, Ed. Academic Press, New York, 312.
31.   Gianturco, M. A.; Giammarino, A. S.; Friedel, Nature 1966, 210, 1358.
32.   Nunomura, N.; Sakai, M.; Asano, Y.; Yokotsuka, T. Agric. Biol. Chem. 1976, 40, 485.
33.   Hodge, J. E. J. Agric. Food Chem. 1953, 1, 928.

RECEIVED June 24, 1985

The Use of High-Performance Liquid Chromatography in Flavor Studies

R. L. Rouseff

Scientific Research Department, Florida Department of Citrus, Lake Alfred, FL 33850

A review of HPLC techniques which have been
used to study flavors in a wide variety of
foods and flavorings is presented. Specific
examples of how HPLC has been used to study
nonvolatile flavor compounds in beer, beef,
tea, grapefruit juice, vanilla, soy bean etc.,
are presented. Relative advantages and dis-
advantages of using HPLC vs GLC are discussed.
Methods to separate specific classes of compounds such as sugars, artificial sweeteners,
bitter/astringent compounds etc.are presented.
Sample pretreatment, sample compatability,
along with examples of auxiliary methods of
peak identification and sensory evaluation
are discussed. Coupled techniques such as
LC/HPLC, and HPLC/GLC/MS systems are
discussed using specific examples.

High performance liquid chromatography, HPLC, has been used
for the separation and identification of a wide variety of flavor
compounds in foods, beverages, flavorings and fragrances. HPLC is
usually but not exclusively used to determine highly polar and/or
thermally unstable compounds. Capillary GLC was and still is the
method of choice to determine volatiles because of the extremely
high resolution and high peak capacity which can be achieved. GLC
detection limits are also in general much lower than those of HPLC.
Therefore, flavor chemists have utilized GLC almost exclusively in
studies involving aroma. Another reason why more GLC than HPLC
flavor studies have been done is because GLC is easily coupled to a
mass spectrograph, which can be used for rapid identification of the
chromatographic peaks. Coupled GC/MS systems have become the
workhorse tool for most flavor chemists. The coupling of HPLC/MS is
much more difficult because of the necessity of removing vast
amounts of solvent before the analytes enter the MS. However, the
coupling of HPLC with MS is an active research area and much
progress has already been made.

0097–6156/85/0289–0079$06.00/0
© 1985 American Chemical Society

In the area of taste, many flavor-active compounds are non-volatile and/or thermally unstable. Since most HPLC separations occur at room temperature, the probability of producing and analyzing artifacts from thermal degradation is greatly reduced. Another major advantage of HPLC is in the area of sample compatability. Since a substance must be dissolved before it can be tasted and since the dissolving liquid is often water, liquid chromatography has the advantage of offering a direct analysis of a flavor-active mixture. HPLC can readily accommodate aqueous samples because the chromatographic mobile phase will generally contain water. GLC systems for the most part do not tolerate water. Therefore, solvent extractions with immiscible organic solvents are required. There are some flavor active compounds such as pyrazines (1) which are difficult to extract from aqueous phases for subsequent GLC analysis. It is often possible to directly inject aqueous samples into HPLC systems, thus circumventing the problem of extracting compounds with low distribution coefficients.

HPLC fractions can often be collected for later sensory evaluation. The objective is to obtain sensory data for each chromatographic peak and to determine which compounds are flavor active. Sniffer ports were used in GLC for the same reason but the sensory evaluation had to be done "on-the-fly" with different sensations coming in rapid succession of each other. Sensory fatigue was often a problem. In contrast, sensory evaluations of HPLC fractions can be obtained under controlled conditions with adequate spacings between samples. However, if sensory evaluation is contemplated, particular care must be taken in choosing the solvents used in the chromatography. Buffers should generally be avoided as well as solvents such as acetonitrile which may pose a health risk. In practice this usually limits the chromatographic solvent system to water and ethanol.

There are many situations where HPLC can be used to advantage in the study of flavors. It is the purpose of this paper to show how HPLC can and has been used to its fullest advantage in studying flavor and to show how HPLC can be used in conjunction with GLC to provide a powerful separations approach.

Sweetness

Sugars Sweetness is one of the most important flavor attributes. Sugars play a major role on the perceived flavor of foods, beverages and flavorings we consume. Because the relative sweetness of individual sugars vary, the flavor chemist needs to know the sugar profile of the product of interest. In order to characterize the raw materials that go into a product or to characterize the final product it is necessary to know the kinds and absolute amounts of individual component sugars. There are a number of classical colorimetric methods to determine the amounts of reducing sugars. Total sugars are determined after acid or enzymic hydrolysis and nonreducing sugars are determined from the difference. Paper and thin-layer chromatography have been used to determine sugars but analysis times are often long, precision poor or quantitation difficult.

Sugars cannot be directly determined by GLC because they melt with decomposition when heated. To circumvent this problem volatile

derivatives can be formed which are heat stable. To minimize
fluctuations in reaction yields an internal standard is usually
required. However, in addition to the inconvenience of an extra
derivatization step, GLC methods suffer from the problem of multiple
reaction products which can produce multiple peaks from a single
sugar.

Except for filtering, most HPLC sugar procedures require
virtually no sample preparation. Therefore, HPLC has quickly become
the method of choice for determining sugars. There are three major
chromatographic approaches for separating sugars. They are:
formation of borate-carbohydrate anions and separation on strong
anion-exchange columns ($\underline{2}$, $\underline{3}$); amino- bonded silica sorption
chromatography using primarily acetonitrile as the mobile phase ($\underline{4}$,
$\underline{5}$); and strong cation-exchange polymers using water as the mobile
phase ($\underline{6}$, $\underline{7}$). Due to advantages in simplicity and reproducibility
most researchers choose one of the latter two approaches. Cation-
exchange systems employ only water as the mobile phase and, thus,
have certain cost and safety advantages over the amino system which
employs primarily acetonitrile. The two systems have opposite
elution orders, so often the choice of which system to use is based
on the relative amounts and types of sugars which need to be
analyzed.

Differential refractometers (RI detectors) have traditionally
been used to measure sugar concentrations as they elute from the
column. However, these detectors are very insensitive, non-specific
and require that only isocratic separations be employed. To remedy
this problem numerous post column reaction (PCR) systems have been
developed ($\underline{8}$-$\underline{10}$). These systems are highly sensitive and highly
specific, however, they can only detect reducing sugars. Further,
PCR systems require additional equipment such as extra pumps and
heated reaction chambers. One of the more promising detectors is
the recently developed pulsed amperometric detector ($\underline{11}$). This
detector is highly specific and highly sensitive. Sugars (reducing
and non-reducing) can now be detected at the ppm level. Several
investigators ($\underline{12}$, $\underline{13}$) have used this detector to determine low
level sugars and sugar alcohols in a variety of products. Shown in
Figure 1 is the sugar chromatogram from flavored potato chips.

<u>Nonsugar sweeteners</u> Artifical sweeteners such as saccharin and
aspertame are readily analyzed using HPLC with minimum sample
treatment. Many low calorie soft drinks contain both artificial
sweeteners because of cost and storage stability advantages. The
taste of aspertame is closest to that of sugar and is thus generally
preferred. However aspertame is more costly and hydrolyzes at the
low pH of soft drinks. Recently Tsang et al. (14) developed a method
to quantify both artifical sweeteners in soft drinks. The reverse
phase method was also used to separate and identify four storage
breakdown products.   Aspertame is quite reactive with aldehydes,
which are the principal flavor compounds in some soft drinks. Thus
the use of HPLC to determine the stability and flavor effects in low
calorie soft drinks will be essential for flavor chemists working in
this area.

HPLC has also been used in the search for other noncaloric
sugar substitues, since many of these substances exhibit many of the
same determination problems as sugars. An example is the recent

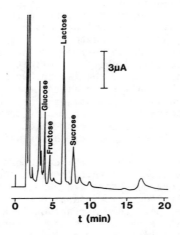

Figure 1. Extract from flavored potato chips. Sugars separated as anions by ion exchange chromatography with a sodium hydroxide eluent and a pulsed amperometric detector. Lactose concentration in the extract is 70 ppm. "Reproduced with permission from Ref. 13 Copyright 1983, 'Marcel Dekker'."

work by Makapugay et al (15) in determining the intensely sweet
triterpene glycoside, mogroside V in Lo Han Kuo fruit from China.
The isolation and identification of the sweet component of this
fruit is quite recent (16) and no previous analytical methods had
been reported. Thus if this compound shows promise as a sugar
substitute HPLC will play an important role in determining the
purity of solutions used in taste threshold tests, levels in
formulations and its storage stability.

Bitterness

Bitterness is an interesting flavor. It can be caused by a host of
structurally dissimilar compounds. On one hand there is a natural
revulsion of things bitter perhaps due to a natural defense
mechanism as many natural toxins are bitter. On the other hand a
limited amount of bitterness is often desirable to produce a clean
taste or something with a bit of "bite." Therefore, it has become
necessary to be able to monitor bitterness by determining which
compounds are bitter and developing methods to quantify those
compounds. HPLC has been used extensively in achieving these twin
goals. It can readily fractionate a product allowing one to
organoleptically determine which fraction is bitter and finally to
separate and identify the specific bitter compound(s).

Hop bitter acids. Hops are used in the brewing of beer to provide
bitterness and aroma. Hop components also contribute to flavor and
foam stability. Humulimic acids from hops are the components
largely responsible for the bitterness of beer (17). A variety of
chromatographic approaches have been proposed for the analysis of
these compounds, they include: ion exchange (18, 19), normal phase
based on silica (20) , paired ion reverse phase (21, 22) and reverse
phase (23, 24). They have been used to evaluate the quality and
quantity of beer bitterness. Correlations between organoletic
bitterness of individual components and the bitterness of beer to
the amounts of these components in beer have been established (25).
HPLC has been used to assess the freshness of hops and to
standardize the use of hop extracts in beer. Precise, specific
analytical techniques such as HPLC are required to determine the
effects of newer processing techniques. In studying super critical
fluid extractions of hops with CO_2, Anderson (26) used HPLC to
monitor the bitter hop acids.

Amino acids and peptides. Several amino acids and peptides are
bitter. These compounds are important flavor components in their
own right and have been implicated as precursors for other
flavor-active reaction products. Traditionally they are separated
using ion-exchange chromatography. Since they do not contain a
strong chromaphore, they are derivatized after column separation to
form products which can be easily detected and quantified. Many
derivatizing agents have been developed. Spackman and coworkers
(27) used ninhydrin and this is still the post column chemistry
employed by most commercial amino acid analyzers. With the advent
of HPLC more efficient separations (28, 29) have been developed
which greatly reduced analysis time. Recently, fluorometric
reagents such as o-phthaladehyde (OPA) (30) have been used to

increase sensitivity. Unfortunately, OPA reacts only with primary amines. Therefore, the addition of oxidizing agents such as sodium hypochlorite (31) is necessary to produce primary amines from secondary amines such as proline.

Several precolumn derivatization techniques are available for those who wish to trade extra sample preparation time for the expense and maintenance of post column pumps and reactors. The more popular derivatives are dansyl-(32), OPA-(33), PTH-(34) and PITC-(35) amino acids. There are problems and limitations with some of these systems, however, analysis time is only 15-25 min. compared to 90-240 min. of the ion-exchange post column systems.

Unfortunately, most flavor researchers have been slow to utilize the speed and sensitivity that HPLC offers. Although bitter amino acids and peptides have been reported in cheese and casein (36, 37) and in yoguart (38) all the analyses were done by traditional methods.

Citrus limonoids and flavanone glycosides. Limonoids are C_{26} fused ring aliphatic compounds containing a furan ring and two lactone rings at opposite ends of the molecule. Limonoids are found in all citrus cultivars. The limonoid in greatest abundance is usually the bitter limonin. Other limonoids are found in lower concentrations and for the most part are non bitter. Bitterness has been a major flavor problem in the citrus industry for years. Thus it has beome necessary to develop sensitive analyses which could separate and quantify the bitter limonoids from the complex citrus matrix. Since limonoids are thermally unstable, HPLC is the technique of choice and several HPLC procedures have been developed (39,40). The procedure probably in most widespread use (41) utilizes a cyano column in the normal phase mode. The bitter limonin and nomilin are well resolved from the nonbitter obacunone and deoxylimonin and other citrus components. However, even though the actual chromatographic analysis is relatively rapid the overall analysis is very time consuming because liquid-liquid extractions are required and the sample preparation is lengthy. A direct injection reverse phase HPLC procedure (42) has been recently developed which greatly reduces overall analysis time as no solvent extraction is required.

Flavanone glycosides are C_{15} heterocyclic ring systems with a disaccharide consisting of glucose and rhamnose attached at the 7 position. These compounds are bitter only if the rhamnose is attached 1-2 to the glucose and nonbitter if attached 1-6. To further complicate the problem not all the bitter forms are equally bitter. In most citrus cultivars either the bitter or nonbitter form are exclusively present. Grapefruit is an exception to this general rule as both forms are present in significant amounts. Fisher and Wheaton (43) developed an isocratic procedure for the separation of the bitter naringin from its non bitter isomer, narirutin. Since at least two other bitter-non bitter isomeric pairs are known to exist in grapefruit a gradient separation has been developed (44). The bitter naringin and the nonbitter narirutin are the major flavanone glycosides in grapefruit and are well resolved. The non bitter hesperidin is well separated from the bitter neohesperidin but hesperidin will sometimes appear as a shoulder of the much larger naringin peak as column efficiency decreases with age. The separation between the bitter poncirin and the nonbitter didymin is not as good but adequate. All the bitter flavanone glycosides in grapefruit juice can be quantified and their relative contribution to juice bitterness can be estimated.

HPLC Analysis of Flavors and Essential Oils

The spicy hot flavor of red peppers is due to capsaicin and related capsaicinoids. Spice manufacturers seek to produce flavorings of consistent flavoring strength. In the case of ground red pepper spice it is necessary to quantify the major red pepper heat principles. An estimate of the spices' "hotness" can then be estimated in a nonsubjective manner as the relative sensory "hotness" for each capsaicinoid has already been established. Since capsaicinoids cannot be determined using GLC without derivatization, several HPLC methods have been developed (45,46) which are based on underivatized direct injections. One of the most recent methods (47)is based on the use of a reverse phase C-18 column with common solvents in the isocratic mode. All the major capsaicinoids could be separated from red pepper extracts in 25 min. To aid laboratories which do not possess capaicinoid standards the authors determined capaicinoid response times and concentration response factors relative to a commercially available material of similiar structure,N-vanillyl-n-nonamide. Since many flavor active compounds are not commercially available, the use of commercially available standards of similar structure in analytical methods should be encouraged. Thus this method is well suited for routine analysis.
      Vanilla extract is used to flavor a wide variety of foods and is very expensive. Herrman and Stockli (48) developed an HPLC method to determine the major components of vanilla extracts. This method can be used to help differentiate natural vanilla extracts from cheaper substitutes. Shown in Figure 2a is a separation of standards of the major components in vanilla. Shown in Figure 2b is the chromatogram from a typical vanilla extract. It should be noted in Figure 2b that natural vanilla extracts apparently lack ethyl vanillin, a commonly used flavor substitute.
      Extracts of licorice root have also been used as a flavoring agent. Hiraga et al. (49) among others have developed an HPLC separation for the constituents in licorice. Because the numerous components in licorice vary considerably in polarity, a reverse phase C-18 column with a water/acetonitrile gradient was employed. Nineteen components were identified. However, a three hour extraction coupled to the 60 min. analysis time probably makes this procedure too lengthy for routine use.
      If only a few specific components need be quantified, the analysis time can be considerably shortened. Dong and DiCesare (50) developed an high speed isocratic reverse phase separation for 2,5-dimethyl-4-hydroxy-3(2H)-furanone. This compound is commonly known as pineapple ketone and can be quantified in food flavoring products using one of the new 3 x 3, C-18 columns in as little as 2 minutes. Other workers (51) have used rapid gradients (6 min.) to separate caffeine, trigonelle and theobromine in coffee and cocoa products.
      It has often been stated that HPLC lacks the resolving power to produce a detailed separation for complex mixture such as

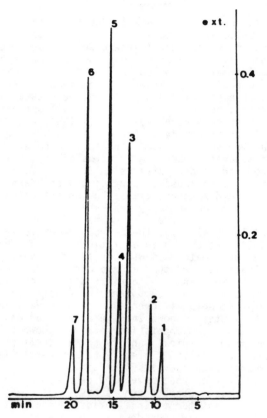

Figure 2a. Chromatogram from 10ul of a standard mixture. Peaks:
1 = 2.5ug of 4-hydroxybenzyl alcohol; 2 = 2.5ug of vanillyl
alcohol; 3 = 2.5ug of 4-hydroxybenzoic acid; 4 = 0.5ug of
4-hydroxybenzaldehyde; 5 = 2.5ug of vanillin; 6 = 2.5ug of ethyl
vanillin; 7 = 0.5ug of coumarin.

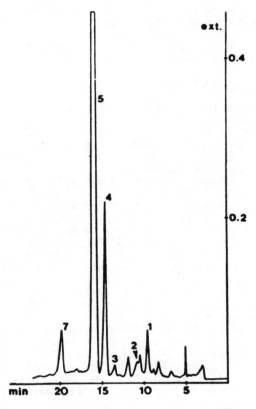

Figure 2b.   Chromatogram from 10ul of vanilla extract.
"Reproduced with permission from Ref. 48 Copyright 1980,
'Elsevier'."

essential oils. The chromatograms shown in Figure 3 developed by
Thies (52) for lavender and peppermint oil are typical. Usually
only 2-3 components are separated to produce a "fingerprint" which
can be used to characterize the product. However, Scott and Kucera
(53) have demonstrated in principle that high efficiency microbore
columns can produce separations comparable to capillary GLC
separations. Shown in Figure 4 is one such separation for cinnamon
essential oil. This detailed separation was achieved with a 1mm
column 14 meters long. However, the separation took over 20 h to
complete which is unrealistically long. Therefore, at least in
terms of the present technology, detailed separations of essential
oils are probably best done using capillary GLC.

HPLC as a Flavor Research Tool

If one is interested in a specific compound or group of related
compounds, HPLC is an excellent tool to study flavor changes in
different products or from different processes. HPLC is generally
regarded as possessing very limited ability to separate large
numbers of compounds with considerably different polarities. Thus
most HPLC procedure determine compounds of a rather narrow polarity
range. This is a disadvantage when studying flavor systems because
one often does not know what is present, or what flavor active
compounds might be formed. However, Qureshi et al.(1,54) have shown
that HPLC can be used to successively and systematically separate a
large number of flavor active compounds of very different polarity.
They developed an HPLC procedure utilizing a single C18 column with
a series of isocratic or gradient runs to successively separate
purines, pyrimidines, nucleosides, nucleotides, polyphenols and
pyrazines in beer. Using this system they were able to quantify 58
compounds in worts and beers at five different stages of brewing.
About 30 additional peaks were resolved but not identified. Beers
using malts form four different barleys had different quercetin to
catechin (Q/C) ratios. Shown in Table I are the polyphenol values
for two canned beers (one American and one English) and a bottled
American beer. Whereas the Q/C ratio is relatively similar for the
two American beers (2.3 vs 3.1) the ratio was considerably lower in
the English beer. Other compounds such as kojic acid, ferulic acid,
p-coumaric acid and cinnamic acid also showed considerable
differences.
    Unfortunately the relative contribution of each of the 58
compounds to beer flavor is known only for a few. There are also
many compounds which were separated by not identified which might be
flavor active. Additional work needs to be done to determine the
relative contribution of the identified compounds to beer flavor and
to identify those unknown compounds which are flavor active.
    In another thorough flavor study Wu et al.(55) determined both
the volatile and nonvolatile flavor compounds found in mushroom
blanching water. They used HPLC to determine such non volatile
flavor components as sugars, amino acids and nucleotides. The free
amino acids were analyzed to determine if they might be involved in
any thermal reactions which might produce Amadori compounds or
Strecker aldehydes which in turn would produce aroma components.
The authors were interested in evaluating three different processing
techniques to recover the flavors in mushroom blanching water.

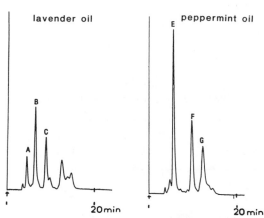

Figure 3. Chromatogram of lavender and peppermint oils. Peaks:
A and E = unknowns, B = linalool, C = linalylacetate, G = methyl-
acetate. "Reproduced with permission from Ref. 52 Copyright 1984,
'Spinger-Verlag'."

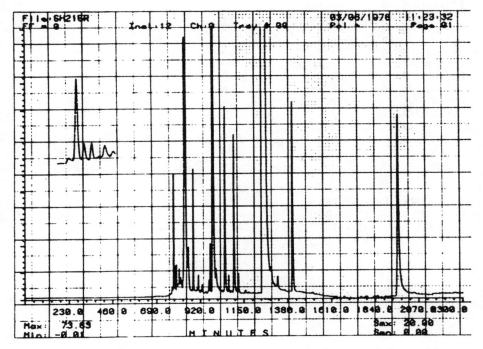

Figure 4. Chromatogram of a cinnamon essential oil. Column, 14m
x 1mm I.D.; solvent ethyl acetate: heptane (5:95); efficiency =
510,000 theoretical plates. "Reproduced with permission from Ref.
53 Copyright 1979,'Elsevier'."

Table I. Polyphenols in Commercial Beers

Compound	Retention Time (min)	Concentration (ug/mL)		
		CE	CA	BA
Koijic acid	4.9	78.2	29.8	28.6
Gallic acid	6.5	20.8	29.2	26.8
3,4-Dihydroxy benzoic acid	10.4	12.8	19.0	18.4
Quercetin	14.2	36.4	125.0	80.8
p-hydroxy benzoic acid	16.3	15.9	12.0	12.0
Catechin	19.4	53.0	55.0	26.0
Vanillic acid	21.9	0.8	0.5	0.3
Ferulic acid	24.1	20.8	4.5	2.2
Caffeic acid	25.4	8.0	7.0	2.0
Gentisic acid	26.9	3.0	2.8	4.6
Syringic acid	27.8	7.0	5.0	3.5
Chlorogenic acid	28.9	4.0	2.0	3.4
Epicatechin	30.5	24.0	13.8	8.5
p-Coumaric acid	36.4	1.6	6.0	7.0
Rutin	42.9	2.7	6.0	6.0
Salicylic acid	43.9	3.1	1.1	1.0
Sinapic acid	48.2	3.9	0.7	0.7
Cinnamic acid	72.5	26.1	3.3	3.5

All values calculated from data obtained at 280 nm.
CE = canned English beer
CA = canned American beer
BA = bottled American beer
Data from Ref. 1.

Glucose was the only major sugar and IMP and GMP were the only major
nucleotides found. A sensory evaluation of the different processed
products indicated a preference for the drum dried product over the
freeze or spray dried product. This preference could not be
explained from sugar or nucleotide values and the amino acid data
was inconclusive. Since the authors have amassed such a large data
pool for both volatile and nonvolatile compounds it is unfortunate
that some form of data analysis such as multivarient statistical
analysis was not applied so as to determine which compounds were
primarily responsible for the perceived flavor preference.

Aqueous tea extracts deteriorate rapidly during storage. They
develop an undesireable taste commonly described as "flat" and
concurrently turn dark brown in color. Roberts et al. (56) used a
modification of the HPLC procedure of Hoefler and Coggon (57) to
separate the polyphenols in tea to study this problem. The effluent
from the reverse phase acetone gradient was monitored at 380 nm.
This wavelength was chosen so that only oxidized polyphenols and no
unoxidized polyphenols would be detected. Thirteen peaks were
resolved; ten were identified as theaflavins and three as
thearubigins. In comparing chromatograms of freshly brewed teas
with teas of 48 and 72 h many new peaks were observed in the
thearubigin region. They concluded that the dark color and off
flavor were due to the formation of oxidized and polymerized
polyphenolic derivatives of the thearubigin type. In a reasonable
but still speculative conclusion they reasoned that the thearubigens
were formed from the unoxidized and partially oxidized compounds
which contribute greatly to the "brisk" taste of tea. Thus the loss
of these compounds upon standing produced a "flat" tasting beverage.
It would appear that the authors have indirect evidence to support
their hypothesis but would need to devise a way to identify and
monitor the loss of the compounds responsible for the "brisk" taste
in order to strengthen their hypothesis.

LC/HPLC

When dealing with complex flavor systems there are certain
advantages to carrying out preliminary liquid chromatographic
separations to obtain several less complex fractions. This allows
one to begin with relatively large amounts of sample. Large amounts
of sample are often required when identifying trace components.
High performance preparative liquid chromatography can often be
employed to rapidly fractionate the sample. Maximum separation
efficiency can be obtained if different chromatographic systems are
used in the preparative and analytical separations. For example,
normal phase chromatography could be used for the initial
fractionization and reverse phase chromatography used for the
anlaytical separation.

Kubeczka (58) utilized this general approach to separate a
terpene mixture into oxygenated terpenes, monoterpene hydrocarbons
and sesquiterpene hydrocarbons fractions before a final analytical
separation. He used a C-18 reverse phase system operated in the
semipreparative mode to obtain his preliminary separations. The
sample fractions were then analyzed by liquid-solid chromatography
using deactivated silica gel (4.8% water/n-pentane) at -15 °C.
Eight monoterpenes were well separated within ten min. Using the

same chromatographic system eight sesquiterpenes could be resolved from numerous impurities/trace components in a natural mixture within 18 min. The compound(s) in each peak were further identified using spectroscopic methods. Analytical HPLC results were compared with those from GLC for each fraction and were in good agreement.

Classical liquid-liquid, liquid-solid and fractional crystillization can also be used before preparative and analytical HPLC to concentrate trace flavor active components. Huang et al. (59) used these classical techniques to isolate a nonvolatile bitter/astringent fraction from soy flour. The bitter fraction was further separated using preparative HPLC. Seven peak fractions were separated and the flavor characteristics of each collected peak determined. The reverse phase analytical separation of the bitter peak fraction produced six well resolved peaks. These peaks were further collected and identified using UV and NMR spectroscopy. Only three of these peaks were bitter. The three bitter peak components were identified as specific isoflavones.

Often when dealing with flavor mixtures containing compounds with large molecular weight differences, gel permentation, GPC, (size exclusion) chromatography can be used to separate the sample into simpler fractions. An example of this approach is the work of Alabran (60) in isolating and identifying beef flavor precursors in individual GPC fractions from raw beef. HPLC was used to separate and identify the individual components. The compounds were further identified using IR and NMR spectroscopy.

HPLC/GLC/MS

The use of HPLC need not be limited to the analysis of nonvolatile flavor components. HPLC can be coupled with GC/MS to form an extremely powerful separation system. The unique separation properties of HPLC can be used to separate complex flavor mixtures to simplify later GC/MS separation and identification. This approach can be somewhat tedious to use in that a separate GC/MS run must be made for each HPLC fraction - but for very complex flavor systems there is often no other choice. Grob et al. (61) devised a direct transfer for HPLC fractions into a capillary GC to make this approach more convenient to use. They demonstrated the practical workability of the system by separating and identifying two components in a toothpaste using both HPLC/GC and HPLC/GC/MS.

In a very detailed study ter Heide and coworkers (62) used HPLC to fractionate a rum flavor extract into 57 components. Each of these components were further analyzed by capillary GC/MS. Over 400 compounds were identified of which 214 had never been reported before as occuring in rum. Thus without the separation power from the combination of HPLC/GC/MS this complex flavor system could not have been chemically characterized to the degree it has in this work.

Yamaguchi et al. (63) used the combination of HPLC/GC/MS to separate the components in a model flavor system. However after HPLC fractionization and before GC/MS analysis each fraction was organoleptically analyzed. There are probably many other flavor systems which could benefit from analysis using the combination of HPLC/GC/MS.

Literature Cited

1. Qureshi, A. A.; Burger, W. C.; Prentice, M. J. Am. Soc. Brew. Chem. 1979, 37, 153-60.
2. Mopper, K. Anal. Biochem. 1973, 56, 440.
3. Mopper, K. Anal. Biochem. 1980, 52, 2018.
4. Schwarzenbach, R. J. Chromatogr. 1976, 117, 206.
5. Wheals, B. B.; White, P. C. J. Chromatogr. 1979, 176, 421.
6. Goulding, R. W. J. Chromatogr. 1975, 103, 229.
7. Hokse, H. J. Chromatogr. 1980, 189, 98.
8. Mopper, K.; Degens, E. T. Anal. Biochem. 1972, 45, 147.
9. Kidby, D. K.; Davidson, D. J. Anal. Biochem. 1973, 55, 321.
10. Honda, S.; Matsuda, Y.; Takahashi, M.; Kakehi, K.; Ganno, S. Anal. Chem. 1980, 52, 1079.
11. Hughes, S.; Johnson, D. C. Anal. Chem. Acta. 1981, 132, 115.
12. Hughes, S.; Johnson, D. C. J. Agric. Food Chem. 1982, 30, 712.
13. Rocklin, R. D.; Pohl, C. A. J. Liq. Chromatogr. 1983, 6, 217.
14. Tsang, W. S.; Clarke, M. A.; Parrish, F. W. J. Agric. Food Chem. 1985, 33, 734.
15. Hussein, M. M.; D'Amelia, R.P.; Manz, A. L.; Jacin, H.; Chen, W.T.C. J. Food Sci. 1984, 49, 520.
16. Makapugay, H. C.; Nanayakkara, N. P. D.; Soejarto, D. D.; Kinghorn, A. D. J. Agric Food Chem. 1985, 33, 348.
17. Takemoto,T; Arihara, S.; Nakajima, T.; Okuhira, M. Yakugaku Zasshi 1983, 103, 1151.
18. Schulze, W. G.; Ting, P. L.; Goldstein, H. In "Liquid Chromatographic Analysis of Foods and Beverages"; Charalambous, G., Ed.; Academic: New York, 1977; Vol. 2, p. 442.
19. Siebert, K. J. J. Am. Soc. Brew. Chem. 1976, 34, 79-90.
20. Slotema, F. P.; Verhagen, L. C.; Verzela, M. Brauwissenschaft 1977, 30, 145-9.
21. Schwarzenbach, R. J. Am. Soc. Brew. Chem. 1979, 37, 180-4.
22. Whitt, J. T.; Cuzmer, J. J. Am. Soc. Brew. Chem. 1979, 37, 41-6.
23. Verzele, M.; Dewaele, C. J. Am. Soc. Brew. Chem. 1981, 39, 67-9.
24. Verzele, M.; De Potter, M. J. Chromatogr. 1978, 166, 320-6.
25. Schulze, W. G.; Ting, P. L.; Henchel, L. A.; Goldstein, H. J. Am. Soc. Brew. Chem. 1981, 39, 12-15.
26. Kowaka, M.; Kokubo, E. J. Am. Soc. Brew. Chem. 1977, 35, 16-21.
27. Anderson, B. M. J. Chromatogr. 1983, 262, 448-50.
28. Spackman, D. H.; Stein, W. H.; Moore, S. Anal. Chem. 1958, 30, 1190.
29. Benson, J. R.; Hare, P. E. Proc. Natl. Acad. Sci. U.S.A. 1975, 72, 619-22.
30. Klapper, D. G. In "Methods in Protein Sequence Analysis"; Elzinga, M., Ed; Humana Press: Clifton, New Jersey, 1982; p. 509.
31. Roth, M. Anal. Chem. 1971, 43, 880-2.
32. Felix, A. M.; Jerkelson, G. Arch. Biochem. Biophys. 1973, 157, 177-182.

33. Englehart, H.; Asshauer, J.; Neue, U.; Weigand, N. Anal. Chem. 1974, 46, 336.
34. Haag, A.; Langer, K. Chromatographia 1974, 7, 659.
35. Cohen, S. A.; Jarvin, T. L.; Bidlingmeyer, B. A. American Laboratory 1984, 51, 48-59.
36. Lowrie, R. J.; Lawrence, R. C. N. Z. J. Dairy Sci. Technol. 1972, 7, 51-3.
37. Okhrimenko, O. V.; Chebotarev, A. I. Izu Vyssh. Uchebn. Zaved., Pishah, Tekhnol 1975, 6, 33-5.
38. Renz, U.; Puhan, Z. Milchwissenschaft 1975, 30, 265-71.
39. Fisher, J. F. J. Agric. Food Chem. 1973, 23, 1199.
40. Fisher, J. F. J. Agric. Food Chem. 1978, 26, 497.
41. Rouseff, R. L.; Fisher, J. F. Anal. Chem. 1980, 52, 1228.
42. Shaw, P. E.; Wilson, C. W. J. Food Sci. 1984, 49, 1217.
43. Fisher, J. F.; Wheaton, T. A. J. Agric. Food Chem. 1976, 24, 898.
44. Rouseff, R. L. Abstracts 1981 Pittsburgh Conference, 1981, p. 364.
45. Sticher, O.; Soldati, F.; Joshi, R. K. J Chromatogr. 1978, 166, 221.
46. Johnson, E. L.; Majors, R. E.; Werum, L.; Reiche, P. In "Liquid Chromaotgraphic Analysis of Food and Beverages"; Charlambous, G.,Ed.; Academic Press: New York, 1979; Vol. 1, p. 17.
47. Hoffman, P. G.; Lego, M. C.; Galetto, W. G. J. Agric. Food Chem. 1983, 31, 1326.
48. Herrman, A.; Stockli, M. J. Chromatogr. 1982, 246, 313.
49. Hiraga, Y.; Endo, H.; Takahashi, K.; Shibata, S. J. Chromatogr. 1984, 292, 451.
50. Dong, M. W.; DiCesare, J. L. Food Technol. 1983, 37, 58.
51. Trugo, L. C.; Macrae, R.; Dick, J. J. Sci. Food Agric. 1983, 34, 300.
52. Thies, W. Fresenius Z. Anal. Chem. 1984, 318, 249.
53. Scott, R. P. W.; Kucera, P. J. Chromatogr. 1979, 169, 51.
54. Qureshi, A. A.; Prentice, N.; Burger, W. C. J. Chromatogr. 1979, 170, 343.
55. Wu, C. C.; Wu, J. L.; Chen, C. C.; Chou, C. C. In "The Quality of Foods and Beverages"; Charlambous, G.,Ed.; Academic Press: New York, 1981, p.133.
56. Roberts, G. R.; Fernando, R, S. S.; Ekanayake, A. J. Food Sci. Technol. 1981, 18, 118.
57. Hoefler, A. S.; Coggon, P. J. Chromatogr. 1976, 129, 460.
58. Kubeczka, K. H. In "Flavor '81: 3rd Weurman Symposium"; Schreuer, P.,Ed.;Walter de Gruyter: Berlin, 1981; p. 345.
59. Huang, A. S.; Hsieh, O. A.; Chang, S. S. J. Food Sci. 1981, 47, 19.
60. Alabran, D. M. J. Agric. Food Chem. 1982, 30, 486.
61. Grob, K.; Fromlich, D.; Schilling, B.; Neukom, H. P.; Nageli, P. J. Chromatogr. 1984, 295, 55.
62. ter Heide, R.; Schaap, H.; Wobben, H. J.; de Valois, P. J.; Timmer, R. In "The Quality of Foods and Beverages"; Charalambous, G.,Ed.; Academic Press: New York, 1981, p.183.
63. Yamaguchi, K.; Satoru, M.; Akiyoshi, A.; Shibamoto, T. In "Liquid Chromatogragraphic Analysis of Foods and Beverages"; Charalambous, G., Ed.; Academic: New York, 1979; Vol. II, p.303.

RECEIVED September 16, 1985

Capillary Gas Chromatographic Analysis of Volatile Flavor Compounds

G. Takeoka, S. Ebeler, and W. Jennings

Department of Food Science and Technology, University of California, Davis, CA 95616

High resolution fused silica capillary columns were used to examine, and in some cases to compare, flavor and aroma essences obtained by 1) direct headspace injection, 2) simultaneous steam distillation extraction, 3) standard Soxhlet extraction using dichloromethane, and 4) high pressure Soxhlet extraction using liquid CO_2. Developments in smaller- and larger-diameter fused silica open tubular columns, and columns coated with a bonded polyethylene glycol stationary phase were also explored.

Gas chromatography remains our most powerful separation technique; probably because of this, it has become the most widely used of all analytical techniques. The major limitation to its application, of course, is that it is restricted to the separation of compounds which possess (or which can be derivitized to compounds which do possess) suitable vapor pressures, or "volatility". Volatility, however, is a necessary attribute of odorous compounds, and aroma exercises a profound influence on flavor. Hence, it is not surprising to find that gas chromatography has received a great deal of attention from the flavor chemist, and that many of the advances in gas chromatography have been made by those in the flavor field.

The flavor chemist, however, faces a variety of complicating factors. Flavor is the integrated and highly subjective response of an individual to a number of different stimuli, while gas chromatography (at its best) merely differentiates some of the compounds present, only some of which serve as flavor stimuli. Flavor samples are often exceedingly complex, some of the constituents are unstable, and they may be differentially dispersed (via preferential solubility, adsorption, absorption) through heterogeneous matrices containing water, fats and oils, protein sols, and/or non-volatile solutes such as sugars and salts. These vagaries in distribution can disrupt (or re-enforce) synergistic and antagonistic interactions between flavor compounds and can also complicate their isolation for analysis.

0097–6156/85/0289–0095$06.00/0
© 1985 American Chemical Society

Even partial resolution of these very complex samples requires the use of very efficient columns. In general, separation efficiency varies inversely with column capacity. Sniff-testing of the column effluent can be very useful to the flavor chemist, but this demands columns of even larger sample capacity. Hence the flavor chemist has been faced with a difficult choice: 1) better resolution of volatile flavor compounds in amounts too small for their sensory evaluation, or 2) sniff testing of poorly resolved multicomponent peaks.

There have been many advances in the field of gas chromatography, and the rate at which new developments occur has been increasing in a manner that is almost exponential. Among those developments are some that hold special significance for the flavor chemist; these include advances in 1) sample preparation, 2) injection hardware and methodology, and 3) column technology.

Sample Preparation

Discussion of sample preparation will be limited to a brief consideration of three techniques:
1) the direct injection of a headspace sample into a fused silica column with on-column cold trapping;
2) simultaneous steam distillation extraction, with a standard atmospheric pressure Nickerson-Likens unit; and
3) Soxhlet extraction, with primary attention to high pressure extractions performed with liquid CO_2.

Recent developments in the area of headspace sampling (1), with an on-column injector that permits insertion of a fused silica needle directly into a fused silica capillary column, make direct headspace injections much more appealing. A short section of column is suitably chilled, and a 200 uL to 500 uL volume of gaseous headspace is injected directly inside the column, thus avoiding its dilution by the carrier gas normally employed to conduct the sample from the injection chamber to the column (Figure 1). The "cryogenic" or "thermal focus" derived from the negative temperature ramp can be complimented by a "phase ratio focus" (1), achieved via a "retention gap" (2). Because the results are quite reproducible, the sensitivity is high, and the approach is extremely simple, the method has been carried forward into other areas, including the detection of purgeable pollutants in water (3), and arson investigations of fire debris (4).

Extraction has long been a popular isolation technique, but it suffers from several limitations. Among the more serious of these are: 1) attempts to remove appreciable amounts of residual solvent from the extract invariably lead to quantitative changes in the sample; 2) important areas of the chromatogram may be obliterated by solvent peaks; 3) solvent odors complicate the sensory analysis of extracted materials; and 4) the toxicity of many common solvents poses stringent limitations on the use of their extracts, but progress also has been achieved in the area of flavor extraction.

From the standpoint of the food chemist, carbon dioxide is one of the more interesting extraction solvents: it is relatively inexpensive, chemically stable, inert, and non-toxic. In addition,

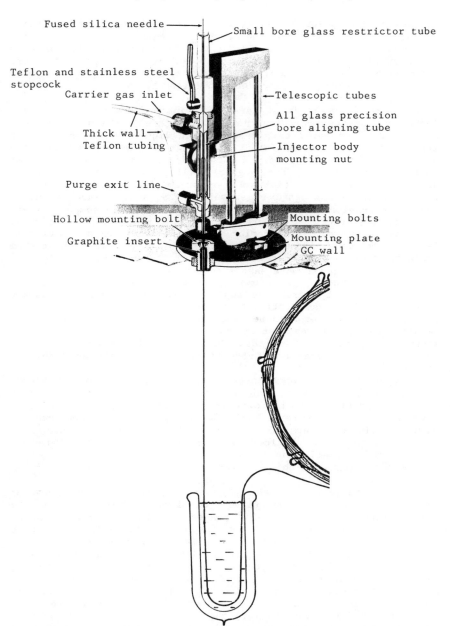

Figure 1. Schematic diagram of a retro-fit on-column injector used for the direct injection of headspace into a fused silica column, by means of a syringe fitted with a fused silica needle. A section of the column is chilled in liquid nitrogen during the injection, the Dewar flask is then removed, and the chromatogram commenced. Adapted from reference (1), which should be consulted for additional details.

its low boiling point facilitates removal of the solvent from the extract. Depending on the temperature-pressure relationships employed, carbon dioxide can be used either as a liquid, or as a supercritical fluid. In the examples explored here, extraction was accomplished by liquid CO_2. The liquid CO_2 was supplied from a standard dip-tube cylinder and purified by passage through an activated charcoal filter.

Figure 2 shows dried tarragon, examined by (top) headspace injection, by (center) simultaneous steam distillation extraction, and (bottom) by Soxhlet extraction with liquid CO_2. The results are reasonable, and consistent with what one might expect: the headspace chromatogram is dominated by the more volatile components, and many of the higher boiling compounds are totally absent. The early portion of the steam distillation-extraction chromatogram is dominated by solvent peaks, which render meaningless this portion of the trace. Some higher boiling components are evident, but their occurence here means that these particular compounds meet another criterion: they must also be steam distillable. In comparison to the others, the bottom chromatogram contains many more peaks; a blank chromatogram (Figure 3) established that all peaks were sample components.

Figure 4 shows the chromatographic results of a black pepper sample examined by (from the top) 1) direct on-column headspace injection, 2) steam distillation-extraction, and 3) Soxhlet extraction with dichloromethane, and with 4) liquid CO_2. Again, the results are quite reasonable, and in good agreement with those shown in Figure 2. Note that CO_2 seems fully as effective (and indeed, in some cases more effective) an extractant as dichloromethane.

While CO_2 is a very interesting extractant from the standpoints of 1) efficiency, and 2) toxicity (i.e., while the extract may cause problems, no toxicity is associated with the solvent or solvent residues, per se), it is also appealing to those dealing with fragile or thermally labile essences. For example many flowers possess very delicate aromas that are easily altered by conventional isolation and concentration treatments. Figures 5-7 show chromatograms of three flower extracts obtained by liquid CO_2 extraction.

New developments in the gas chromatographic columns are of special interest to the flavor chemist. These include the availability of improved polar phases, columns of "non-standard" dimensions (i.e. "microbore" and "megabore" columns), and columns with "super-thick" (i.e. 3 to 5 um) films of stationary phase.

One of the more interesting developments in stationary phases is the availability of "bonded" polyethylene glycol (PEG)-type phases. Most of these bonded PEG phases suffer the same limitations as Carbowax 20-M; they are extremely susceptible to oxygen, are soluble in water and low molecular weight alcohols, and (most important to the flavor chemist) they solidify at a relatively high temperature, rendering them of limited use for the analysis of lower molecular weight solutes. However, the particular bonded PEG used in Figure 8, while still susceptible to oxidative degradation, is less limiting in the other respects: it is firmly bonded, non-extractable with water or methanol, and it can be used at tempera-

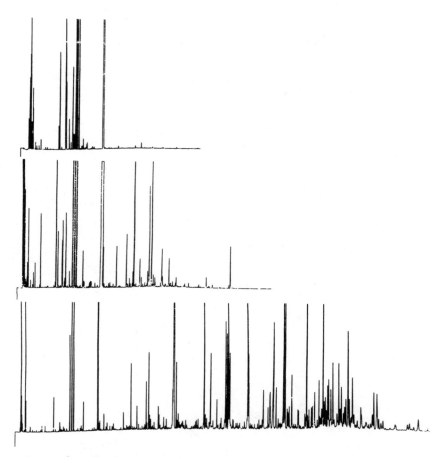

Figure 2. Chromatograms obtained by the application of various sampling techniques to leaves of dried tarragon. Top, 200 uL of headspace vapor in equilibrium with the sample at 45°C, analyzed by direct headspace injection, using the apparatus depicted in Figure 1; center, 1:100 split injection of 1 uL of a steam distillation-extraction essence (pentane) from a standard Nickerson-Likens apparatus*; bottom, 1:100 split injection of 1 uL of an extraction essence obtained with high pressure Soxhlet extraction with liquid CO_2*. Chromatographic conditions, 30 m x 0.25 mm fused silica column coated with a 0.25 um film of bonded polymethylsiloxane (DB-1*), 40°C to 280°C at 4°C/min.

*J and W Scientific, Inc.

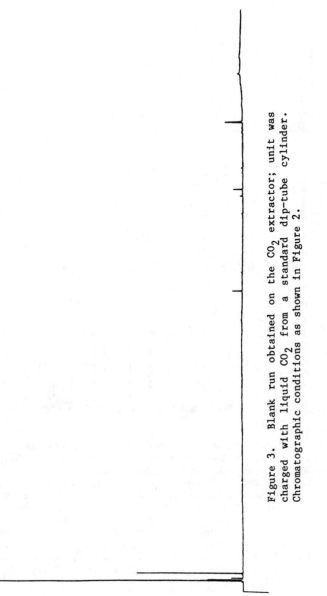

Figure 3. Blank run obtained on the CO_2 extractor; unit was charged with liquid CO_2 from a standard dip-tube cylinder. Chromatographic conditions as shown in Figure 2.

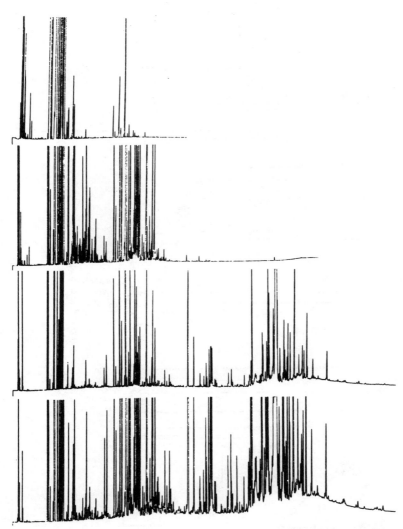

Figure 4. Chromatograms obtained by the application of various sampling techniques to aliquots of a Lampong black pepper sample. Top, 200 uL of headspace vapor; upper center, steam distillation-extraction essence; lower center, Soxhlet extraction with dichloromethane; bottom, high pressure Soxhlet extraction with liquid CO_2. All other conditions as in Figure 2, above.

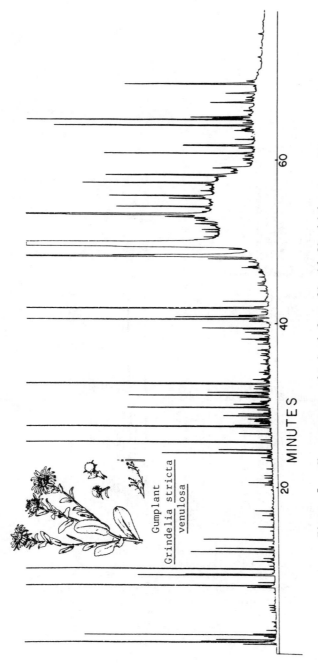

Figure 5. Chromatogram obtained from liquid CO_2 high pressure Soxhlet extraction of Grindelia stricta venulosa (gumplant). Conditions as shown in Figure 2.

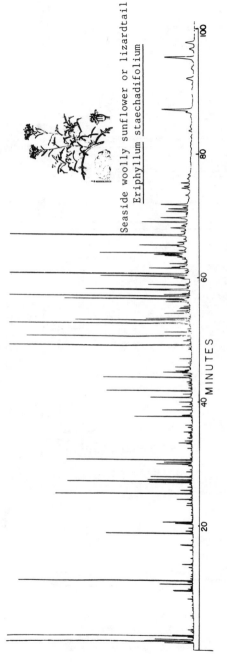

Figure 6. Chromatogram obtained from liquid CO_2 high pressure Soxhlet extraction of Eriophyllum staechadifolium (seaside woolly sunflower, also known as lizardtail). Conditions as shown in Figure 2.

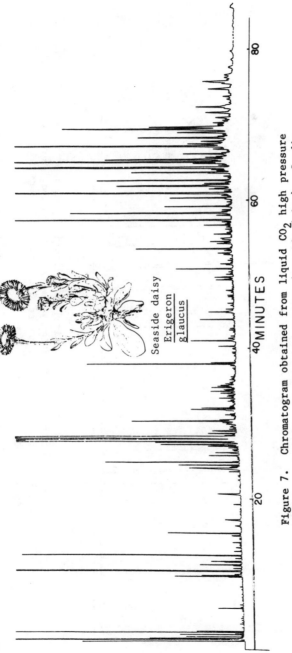

Figure 7. Chromatogram obtained from liquid CO_2 high pressure Soxhlet extraction of Erigeron glaucus (seaside daisy). Conditions as shown in Figure 2.

tures as low as 0°C. As shown in Figure 8, the resistance to water and alcohol of this particular bonded PEG has been exploited to achieve analysis of alcoholic beverages by direct split injection; the low temperature capability permitted a starting temperature of 35°C, and resulted in baseline separation of acetaldehyde, methyl acetate, ethyl acetate and methanol before the appearance of the ethyl alcohol peak. The shape of the latter peak is quite good, and facilitates (rather than interferes with) the assessment of intermediate and higher boiling components. Such injections, however, do result in the deposition of wine solids and other non-volatile residues on the front of the column; in time, this invariably leads to a loss of column efficiency. However, since these phases are non-extractable with water, the water-soluble

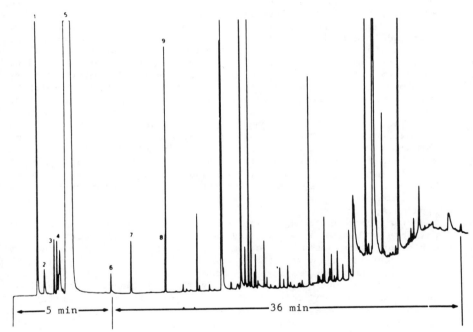

Figure 8. Chromatogram of a German white wine (Mosel region) on a bonded polyethylene glycol phase, capable of low temperature operation. Split injection. Note that by starting at 35°C, acetaldehyde (1), methyl acetate (2), ethyl acetate (3), methanol (4), and ethyl alcohol (5) have all been cleanly resolved. Column, 30 m x 0.25 mm DB-Wax*; 35°C for 5 min, 6°/min to 230°C, 5 min hold. Other peak assignments: (6) 1-propanol; (7) 2-methyl-1-propanol; (8) 2-ethyl-1-butanol; (9)3-methyl-1-butanol.

*J and W Scientific, Inc.

residues can be removed by periodically back-flushing the column with a few mL of distilled water at room temperature.

Other developments in column technology include the availability of columns with smaller and with larger diameters, coated with a wide range of stationary phase film thicknesses (5, 6). Although these topics have always interested chromatographers, recent advances can be largely attributed to (in microbore) Schutjes and his colleagues (e.g. (7-9)), and (in megabore) to Ryder et al. (10). Grob (11, 12) was probably the first to achieve practical results with ultra-thick film columns, which were also considered by Jennings (13, 14), by Sandra et al. (15) and by Ettre (16, 17).

Separation efficiency in terms of the number of theoretical plates per meter of column length varies inversely with column radius: better separation is achieved on smaller diameter columns. Columns whose inner diameters are less than 100 um, however, are extremely difficult to interface with normal inlets and detectors. In addition, their capacities are very limited, they are easily overloaded, and their behaviour with inlet splitters (which at the present time is the most practical means of introducing a sample on these very small bore columns) can be capricious. Even the 100 um ID column suffers from these limitations; skilled chromatographers have used them to good advantage, but at our present state-of-the-art, many will experience considerable frustration with these columns.

The development of the megabore column may go down in chromatographic history as having the most benefit for the most chromatographers. The vast majority of chromatographers have never experienced the thrill of capillary chromatography. Some 75% of the chromatographic community continues to employ packed columns. Many such users accept the limitations of the packed column because they are more comfortable with the more familiar, simpler apparatus, and are apprehensive about making any changes to the injector, wary of "complex" make-up gas adapters, suspicious about possible dilution effects, and unhappy with the facts that a different flame jet may be required, and that they must develop a better understanding of the system. Even the word "capillary" triggers a negative response in these users.

Judged by our present standards of performance, the megabore is not a capillary column; it is a large diameter open-tubular column. It achieves its optimum linear gas velocity at flow volumes of ca. 5–7 cc/min. These velocities are too low to rapidly flush the inlet that is normally present on the packed column instrument, and as a consequence would lead to the slow introduction of a broad, dilute band of sample, resulting in poor separation. Similarly, delivery of this restricted volume of carrier gas to the flame ionization detector would disrupt the "normal" gas flow ratios; the detector would not be operating in a plateau region, and sensitivity would be lower. Alternatively, by sacrificing approximately 50% of its potential separating efficiency, the megabore can be operated at velocities that would satisfy the requirements of the packed column inlet and the detector (e.g. 30 cc/min). Operation under these conditions makes it possible to substitute the megabore directly for the packed column without instrumental modification; even under these "less-

than-ideal" conditions, as compared to the packed column, the much
more inert megabore delivers superior separation in a shorter
analysis time at increased sensitivity (Figure 9).   Obviously,
quantitative reliability will also be enhanced.   Because it can be
directly substituted, the megabore column offers a realistic
alternative to the majority of the chromatographic community who
still use packed columns. If increased resolution is ever required,
the separation efficiency can be doubled by reducing the carrier
flow toward its optimum value, and adding make-up gas at the
detector.

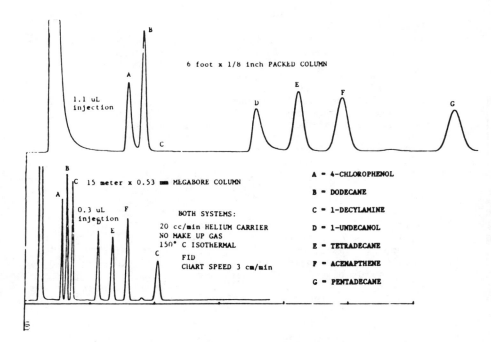

Figure 9.   Results of the direct substitution of a large-bore
fused silica open tubular column.   Solutes (in order of
elution):   4-chlorophenol, dodecane, 1-decylamine, 1-undecanol,
tetradecane, acenaphthene, pentadecane.   Top, 6 ft x 1/8 in OD
stainless steel column packed with Chromosorb W coated with
OV 101; 20 mL/min helium carrier gas; 1.1 uL injected on-
column.   Bottom, same inlet, same detector (FID), 15 m x 0.53
mm Megabore column coated with a 1.5 um film thickness of
bonded polymethylsiloxane (DB-1) substituted, and operated at
the same 20 mL/min helium flow; 0.3 uL injected in same
manner.   The time scale is the same in both chromatograms.
Note tailing of the phenol and alcohol, and complete
disappearance of the amine on the packed column.

Acknowledgments

Portions of Figure 1 were reproduced with permission of J and W Scientific, Inc. The authors are also grateful to R.R. Freeman, E. Guthrie, R. Lautamo, and L. Plotczyk for data on the Megabore column, and for access to the chromatograms shown in Figures 8 and 9.

Literature Cited

1. Takeoka, G.; Jennings, W.   J. Chromatogr. Sci. 1984, 22, 177–184.
2. Grob, K., Jr.; Mueller, R.   J. Chromatogr. 1982, 244, 185–196.
3. Mehran, Mehrzad, Cooper, W. J. and Jennings, W.   Paper No. 21, presented at ACS National Meeting, 28 April–03 May 1985, Miami Beach, FL.
4. Reeve, V.; Jeffries, J.; Weihs, D.; Jennings, W.   Paper presented at Labcon West, 23–25 April 1985, San Mateo, CA.
5. Jennings, W.   Paper presented at the 09 October ASTM Meeting, 09 October 1984, St. Louis, Mo.
6. Jennings, W.; Takeoka, G.   Paper presented at "Neue Ulmer Gespraeche", Symposium on Capillary Chromatographic Analysis of Drugs and Pharmaceuticals, 06–09 May 1985, Neu Ulm, Germany.   Proceedings in press, Huethig Publishing Co.
7. Schutjes, C. P. M.   Doctoral Thesis, Eindhoven University of Technology, The Netherlands, 1983.
8. Schutjes, C. P. M.; Vermeer, E. A.; Cramers, C. A.   Proc. 5th Int. Symp. on Capillary Chromatography, 1983, p. 29.
9. Schutjes, C. P. M.; Vermeer, E. A.; Scherpenzeel, G. J.; Bally, R. W.; Cramers, C. A.   J. Chromatogr. 1984, 289, 157–162.
10. Ryder, B. L.; Phillips, J.; Plotczyk, L. L.; Redstone, M. Paper No. 497, Pittsburgh Conference on Analytical Chemistry and Applied Spectroscopy, 5–9 March, 1984, Atlantic City, NJ.
11. Grob, K., Jr.; Grob, K.   Chromatographia 1977, 10, 250–255.
12. Grob, K.; Grob, G.   J. High Res. Chromatogr. 1983, 6, 133–139.
13. Jennings, W.   Paper No. 286, Pittsburgh Conference on Analytical Chemistry and Applied Spectroscopy, 5–9 March, 1984, Atlantic City, NJ.
14. Jennings, W.   J. Chromatogr. Sci. 1984, 22, 129–135.
15. Sandra, P.; Temmerman, I.; Verstappe, M.   J. High Res. Chromatogr. 1983, 6, 501–504.
16. Ettre, L. S.   Chromatographia 1983, 17, 553–559.
17. Coleman, P.; Ettre, L. S.   J. High Resol. Chromatogr. 1985, 8, 112–118.

RECEIVED August 21, 1985

High-Resolution Gas Chromatography–Fourier Transform IR Spectroscopy in Flavor Analysis
Limits and Perspectives

Heinz Idstein and Peter Schreier

Lehrstuhl für Lebensmittelchemie, Universität Würzburg, Am Hubland, D–8700 Würzburg, Federal Republic of Germany

The capability of high–resolution gas chromatography coupled with Fourier transform infrared spectroscopy (HRGC–FTIR) for the analysis of food flavors is demonstrated by selected examples (tropical fruits, endive, fermentation byproducts). A problem associated with HRGC–FTIR analysis in flavor research is its relative low sensitivity and dynamic range compared to mass spectrometry (MS). To overcome this problem, a dynamic compression ("DYCOM") system consisting of a linked HRGC–FTIR–MS combination, in which multidimensional packed-capillary gas chromatography (MDGC) is integrated, is proposed. With this system, operation with full sensitivity of both spectroscopic methods is possible.

The separation of volatile trace components usually is achieved with high–resolution gas chromatography (HRGC) using capillary columns (1). However, with a complex mixture of substances of different chemical classes, as found among food flavors, one separation technique may not be sufficient to provide the maximum amount of information about the constituents, and repeated separations under a variety of chromatographic parameters may be necessary. More and improved information can be obtained by simultaneous introduction of the sample onto two columns of different polarity and parallel double detection (e.g., a nonspecific detector in tandem with a selective detector) (2). Gas chromatography–mass spectrometry (HRGC–MS) occupies a special place among the analytical techniques for investigations of volatile flavor compounds, since it provides maximum information from minimum sample material (3). HRGC–MS provides both chromatographic (linear retention index) and structure–specific information (MS–spectrum). Nevertheless, the method is not capable of discriminating between different isomers. In these cases, one must obtain additional information with infrared or NMR spectroscopy.

Early attempts to combine an IR instrument with an analytical gas chromatograph were only partly successful; complications were related to the fact that components eluted from the GC column in time intervals too short to permit matching the scale time of the IR

0097–6156/85/0289–0109$06.00/0

equipment then available (4). Recently, dispersive IR units have been modified to make them compatible with GC, and additional sensitivity was gained by applying Fourier transform (FT), leading to both better signal-to-noise ratio, and/or spectra in a shorter time (5-7).

The importance of obtaining IR information in addition to mass spectra will be demonstrated by an example taken from the work of our colleague Adam (8) (Figure 1). After photochemical transformation of compound 1 the main reaction product 2 (> 95 %) (Figure 1, top) was separated by preparative gas chromatography and then characterized by various spectroscopic methods. As to another reaction product (< 5 %) detectable by HRGC (Figure 1, bottom left; black square) and characterized by MS (Figure 1, bottom right) the hypothetical structures 3-8 can be formulated. Due to the low amount of the product, NMR spectroscopy could not be used for structure elucidation, but HRGC-FTIR analysis solved the problem. As outlined in Figure 2 the vapor phase FTIR spectrum clearly indicated the occurrence of an exocyclic methylene group (887 cm^{-1};(9)) excluding all structures except formula 3 (Figure 1).

In the area of flavors, we recently reported the results obtained from HRGC-FTIR studies of tropical fruits (10). During our HRGC-MS studies on cherimoya (Annona cherimolia, Mill.) volatiles, we obtained, among others, the mass spectrum shown in Figure 3. At first glance, the spectrum suggests that the compound could be 2-pentenol, and the unkwown and 2-pentenol had similar GC retention times. However, the "on-the-fly" IR spectrum of the same unknown (Figure 4) showed absorption bands at 3082, 1755, 1650, 1177 and 895 cm^{-1} indicating an unsaturated (Z)-configurated ester. Comparison of the chromatographic and spectroscopic data with those of synthesized reference samples proved that the identity of the unknown compound was (Z)-2-pentenyl butanoate.

Through support provided by various FTIR manufacturers,we were able to extend our HRGC-FTIR studies of volatiles including guava fruit (Psidium guajava, L.) (11), fermentation byproducts (12) and vegetables (13). The following examples are drawn from these investigations. Figure 5 shows the FID trace from the HRGC-FTIR study of a polar silica gel fraction of volatiles (alcohols, hydroxy esters, lactones etc.) obtained from model fermentations using different yeast species and strains. The numbers in circles indicate the registration of FTIR spectra. In some cases, minor constituents were detected by FTIR, since they were good IR absorbers. Some examples of vapor phase FTIR spectra taken from this run are presented in Figure 6, i.e. those of two major components, 2-methyl-1-propanol and 2-phenylethanol (top, left and right, respectively) and of two compounds, ethyl lactate (bottom left) and γ-butyrolactone (bottom, right) that were detectable in much lower concentrations.

Similar results were obtained in our study of a mid-polar silica gel fraction of endive volatiles. Figure 7 shows the FID trace of this fraction and again the numbers in circles indicate where FTIR spectra could be recorded. In this case, practically all major compounds were detected (and identified) by IR spectroscopy, again including minor

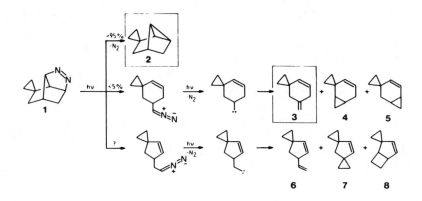

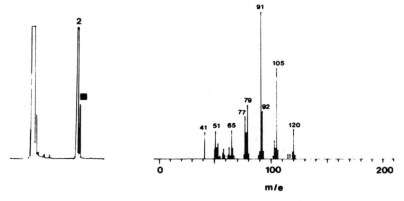

Figure 1. Top: Major product (2) (>95 %) and hypothetical structures (3–8) (<5 %) of photochemical transformation of 1 (8). Bottom left: HRGC separation of the byproduct (black square). Bottom right: Mass spectrum (EI, 70 eV) of the byproduct (8).

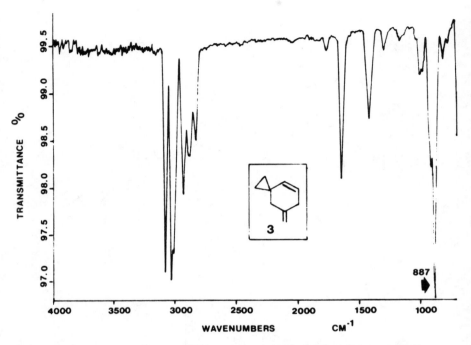

Figure 2. Vapor phase FTIR spectrum of the byproduct of photo-chemical transformation of 1 indicating structure 3 (cf. Figure 1).

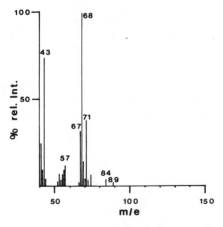

Figure 3. Mass spectrum (70 eV) of unknown compound of cherimoya fruit volatiles (explanation cf. text).

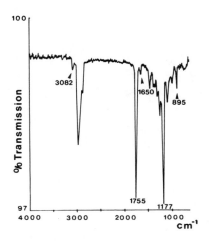

Figure 4. Vapor phase FTIR spectrum of unknown compound from HRGC–FTIR analysis of cherimoya fruit volatiles. Same compound as in Figure 3; explanation cf. text.

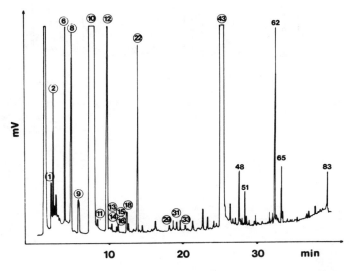

Figure 5. FID trace from the HRGC–FTIR study of silica gel fraction III of fermentation byproducts obtained in model fermentation studies with <u>Saccharomyces</u> <u>cerevisiae</u> (<u>12</u>). Column: 30 m x 0.32 mm i.d. CP 57 CB (df = 0.2) (Chrompack). 50–130° C, 2°/min. 130–200°C, 5°/min.

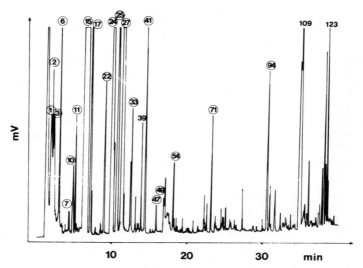

Figure 6. FID trace from the HRGC–FTIR study of silica gel fraction II of endive volatiles (13). Column: 30 m x o.32 mm i.d. CP 57 CB (df = 0.2) (Chrompack). 50–130° C, 2°/min. 130–200° C, 5°/min.

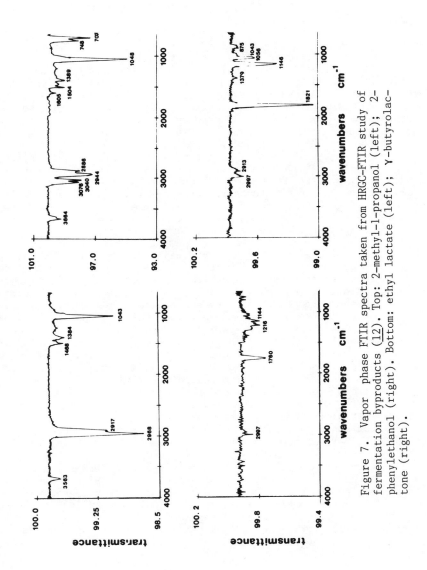

Figure 7. Vapor phase FTIR spectra taken from HRGC–FTIR study of fermentation byproducts (12). Top: 2-methyl-1-propanol (left); 2-phenylethanol (right). Bottom: ethyl lactate (left); γ-butyrolactone (right).

volatiles in some cases. Some vapor phase IR spectra from this run are selected as examples in Figure 8. In this Figure, the "on-the-fly" spectra of (E)-2-hexen-1-ol (top, left), (Z)-3-hexen-1-ol (top, right), 1-penten-3-ol (bottom, left) and phenylethanal (bottom, right) are presented. The IR spectra and the other recorded FTIR data were discussed elswhere (12,13).

As already reported for our recent study on tropical fruits, the principal capability of HRGC-FTIR for the analysis of food flavors employing WCOT columns was confirmed; however, there are some limitations. First of all, the sensitivity of the method has permitted recording HRGC-FTIR spectra for some fractions of as little as about 50 ng/peak. On the other hand, detection of strong bands exhibited by a molecule does not always enable component identification, especially if one considers that a further limitation to rapid and positive identification by HRGC-FTIR analysis is still the lack of adequate IR vapor phase spectral banks (14). Attempts are being made to collect such data (15,16). As to sensitivity, the present state-of-the-art is that on a routine basis the HRGC-FTIR method requires approximately 50 ng/peak, while the GC-MS technique requires less than 1 ng/peak in order to record good interpretable IR- and MS spectra. This indicates that HRGC-FTIR is more than one order of magnitude less sensitive than HRGC-MS for present state-of-the-art systems.

This last-mentioned fact is also the major limitation in coupling both IR and MS in tandem, which is principally possible due to the non-destructive character of IR spectroscopy. Three important papers have been provided on this topic (17-19), but in all three publications the chromatographic procedures and coupling systems did not correspond to the state-of-the-art in this field, i.e. SCOT columns with low separation capacity or even a jet separator were used. Consequently, these combinations were only optimized for FTIR spectroscopy, which led to strong loss in sensitivity of mass spectrometry in comparison to HRGC-MS coupling systems. Whereas with a HRGC-FTIR-MS system problems of overloading arise in HRGC and MS, in FTIR spectroscopy one has to extend the detection limit to a value common for MS. Our idea is to solve this problem by using packed-capillary multidimensional gas chromatography (MDGC) with a suitable gas chromatographic apparatus. Figure 9 outlines our ideas for a linked HRGC-FTIR-MS system, which can obtain the highest sensitivity of both spectroscopic techniques. In this scheme, in which the loading capacity of the column (a), the detection limit of FTIR (b) and that of MS (c) are outlined, four typical examples for the concentration dynamics of peaks (1-4) separated by gas chromatography are represented. In the combination packed GC (PGC)-FTIR-MS (A) peak 1 is eluted in the optimal area for detection with both spectroscopic methods; peak 2 is overloaded and may contain more than one component; peak 3 can only be detected by MS, and the amount of peak 4 is even below the MS detection limit.

In comparison to system (A), using HRGC (B) the separation capacity can be improved, e.g., leading to separation of compounds 1 and 3 to 1a/1b and 3a/3b, respectively, but the dynamics of concentrations are not influenced. With traditional MDGC (B), i.e. using partial "heart-

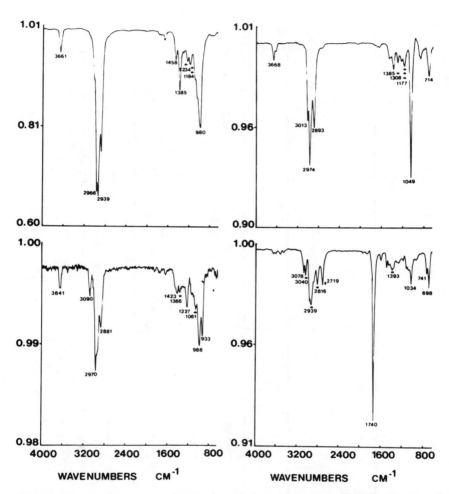

Figure 8. Vapor phase FTIR spectra taken from HRGC–FTIR study of endive volatiles (13). Top: (E)-2-Hexen-1-ol (left); (Z)-3-Hexen-1-ol (right). Bottom: 1-penten-3-ol (left); phenylethanal (right).

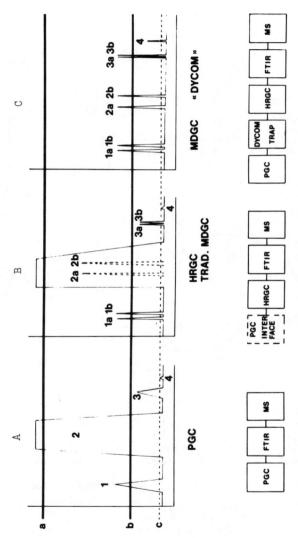

Figure 9. Proposal for a linked HRGC–FTIR–MS system with integrated multidimensional gas chromatography (MDGC) to be used with full sensitivity of both spectroscopic techniques. A: Packed GC–FTIR–MS. B: Traditional MDGC system with FTIR–MS. C: Proposed "DYCOM" system consisting of packed-capillary MDGC–HRGC–FTIR–MS. a: Loading capacity of column. b,c: Detection limits of FTIR and MS, respectively. Further explanations, cf. text.

cut", the amount of the overloaded compound 2 can be reduced in such a way that with HRGC an improved separation will be obtained (2a/2b). Nevertheless, the lack of spectroscopic detection for peaks 3a/3b (FTIR) and 4 (MS) still remains. Our proposal for a dynamic compression of concentrations dynamics as needed for flavor mixtures is outlined in system C. The dynamic compression ("DYCOM") works in the following way: Peaks 1a/1b and 2a/2b would be handled by the traditional MDGC technique, i.e. partial "heart-cut" would be performed, but using the full flow of the packed column onto the "DYCOM" trap, compounds 3a/3b and 4 would be enriched and could then be detected by FTIR (3a/3b) and MS (4), respectively. The individual elements of such a device are all commercially available, e.g., the first three elements of system C practically correspond to the Sichromat-2 MDGC apparatus from Siemens (20). As peak triggering is desirable, the computerization of the commercial system should be extended. Such a linked HRGC-FTIR-MS system using the full sensitivity of both spectroscopic methods has not been described as yet, but we are sure that it will be realized in sophisticated laboratories in the near future.

Acknowledgments

We are indebted to the manufacturers Bruker, Bio-Rad Digilab, and Nicolet for the analysis of flavor mixtures by HRGC-FTIR. Special thank is expressed to Mrs. E.M. Götz-Schmidt and Mr. R. Hock for their contributions to sample preparation.

Literature Cited

1. Jennings, W.; Takeoka, G. In "Analysis of Volatiles"; Schreier, P., Ed.; W. de Gruyter: Berlin, New York, 1984; pp. 63-75.
2. Schomburg, G.; Husmann, H.; Podmaniczky, L.; Weeke, F.; In "Analysis of Volatiles"; Schreier, P., Ed.; W. de Gruyter: Berlin, New York, 1984; pp. 121-150.
3. Ten Noever de Brauw, M.C.; In "FLAVOUR '81"; Schreier, P., Ed.; W. de Gruyter: Berlin, New York, 1981; pp.253-286.
4. Low, M.J.D.; Freeman, S.K.; J. Agric. Food Chem. 1968, 16, 525-528.
5. Ericksson, M.D.; Appl. Spectrosc. Rev. 1979, 15, 261-278.
6. Griffiths, P.R.; De Haseth, J.A.; Azarraga, L.V.; Anal. Chem. 1983, 55, 1361 A.
7. Herres, W.; In "Analysis of Volatiles"; Schreier, P., Ed.; W. de Gruyter: Berlin, New York, 1984; pp. 183-217.
8. Adam, W.; Dörr, M.; Hill, K.H.; Peters, E.M.; Peters, K.; Von Schnering, H.G.; J. Amer. Chem. Soc., 1984, in press.
9. Nyquist, R.A.; "The Interpretation of Vapor-Phase Infrared Spectra. Vol. 1. Group Frequency Data"; Sadtler/Heyden: Philadelphia, 1984.
10. Herres, W.; Idstein, H.; Schreier, P.; HRC & CC 1983, 6, 590-594.
11. Schreier, P.; Idstein, H.; Herres, W.; In "Analysis of Volatiles"; Schreier, P., Ed.; W. de Gruyter: Berlin, New York, 1984; pp. 293-306.
12. Hock, R.; Dissertation, Univ. Würzburg, 1985.
13. Götz-Schmidt, E.M.; Dissertation, Univ. Würzburg, in prep.

14. Welti, D.; "Infra-Red Vapour Spectra"; Heyden: London, 1970.
15. Nyquist, R.A.; "The Interpretation of Vapor-Phase Infrared
     Spectra. Vol. 2."; Sadtler/Heyden: Philadelphia, 1984.
16. "FLATCO" (Flavours-Attractants-Contaminants) HRGC-FTIR-MS   data
     bank; Schreier, P., Ed.; STN International, Columbus, Karlsruhe,
     Tokyo, in prep.
17. Wilkins, C.L.; Giss, G.N.; Brissey, G.M.; Steiner, S.; Anal.
     Chem. 1981, 53, 113-117.
18. Crawford, R.W.; Hirschfeld, T.; Sanborn, R.H.; Wong, C.M.; Anal.
     Chem. 1982, 54, 817-820.
19. Wilkins, C.L.; Giss, G.N.; White, R.L.; Brissey.G.M.; Onyiriuka,
     E.C.; Anal. Chem. 1982, 54, 2260-2264.
20. Oreans, M.; Müller, F.; Leonhard, D.; Heim, A.; In "Analysis of
     Volatiles"; Schreier, P., Ed.; W. de Gruyter: Berlin, New York,
     1984, pp. 171-182.

RECEIVED August 26, 1985

Tandem Mass Spectrometry Applied to the Characterization of Flavor Compounds

Kenneth L. Busch and Kyle J. Kroha

Department of Chemistry, Indiana University, Bloomington, IN 47405

Tandem mass spectrometry (MS/MS) is a new analytical technique applied to problems in food and flavor analyses. Rapidity of analysis, a high discrimination against chemical noise, and the ability to analyze mixtures for functional groups are attributes of MS/MS that make it attractive for such problems. Sample collection and pretreatment differ from methods used in GC/MS. Correct choice of an ionization method is paramount. Daughter ion MS/MS spectra are used for compound identification via comparison with those of authentic compounds, and parent and neutral loss spectra are useful in functional group analysis. Applications to direct analysis of volatiles emitted from fruits and to spice analyses are considered.

Why use MS/MS analysis of volatile components from food and flavor components? Figure 1 provides the answer. The top trace is the capillary column gas chromatographic profile of the concentrated volatiles from a knockwurst sausage sample. The temperature program of 55°C to 180°C at 5° per minute establishes the time scale from beginning to end of run as 25 minutes. Coupled to a mass spectrometer for identification, each of the many compounds can be examined by the mass spectrometer for only a few seconds. The bottom series of figures illustrates the direct MS/MS analysis of the volatiles from the sausage sample. A stream of air is swept over the sausage and carried into the source of the mass spectrometer. Ions are formed from the volatile constituents, and the first analyzer of the instrument scans (5 s) to provide a mass spectrum of the mixture. A particular ion is selected from all of those formed, excited by collision, and its fragment ions analyzed by a second mass analyzer (5 s). The MS/MS spectrum thus obtained is compared to the spectrum of the authentic compound (contained in the laboratory library) and the identity of the compound established (10 s).

The total time for analysis by MS/MS is under a minute, including the time required to load the sausage into the sample

0097–6156/85/0289–0121$06.00/0

boat. A particular ionic component present in the mass spectrum can be identified in only a few seconds. Since all ions are available continuously, the acquisition of data can be tailored to the intensity of the signal. For strong signals, a few seconds suffices; for weak signals, the integration time can be lengthened appropriately. Note that the ion chosen for the MS/MS experiment at m/z 163 is only a minor component in the mass spectrum. The analyst has the freedom to examine any ion formed in the source in any order, unlike GC/MS, which allows examination of the sample only in a short "time window" established by the chromatography. To re-examine a compound in GC/MS, the entire sample must be reintroduced to the gas chromatograph. In MS/MS, the original ion is simply again selected by the first mass analyzer. Finally, the time evolution of a number of compounds can be followed directly with MS/MS, for example, as the sample is heated. In GC/MS this simple experiment generates a number of samples, each of which must be discretely analyzed.

Background. As an analytical technique, tandem mass spectrometry is just entering its second decade of development. The variety of reported applications belies its relative youth. Tandem mass spectrometry (MS/MS) grew out of early work which used metastable ion transitions in order to establish ion structures and interrelationships. After extensive applications to ion structural studies, its usefulness in direct complex mixture analysis became apparent with the early work of Cooks (1-3). Its successes in problem solving are summarized in a recent book edited by McLafferty (4). Now, with several commercial instruments available, MS/MS is being evaluated for application in several new areas, including biochemical analysis, forensic chemistry, and food and flavor analyses. The principles of MS/MS will be summarized in the first part of this chapter. The second part of the chapter will deal with the reported applications of MS/MS to flavor analysis.

Principles

Ion Processing. As mentioned, MS/MS began with the study of metastable ions (5). Metastable transitions are observed from ions which undergo a dissociation while in transit through the instrument. The transition is a chemical reaction characteristic of the nature of the ion. In MS/MS, the instrument is modified so that the reactions occur more frequently and the masses of the reacting ion and the product ion can be established.

The approach to MS/MS is thus quite different from that for high resolution mass spectrometry. There the exact mass measurement which provides the empirical formula of the ion is a physical measurement. In metastable ion studies, the focus is on the nature of the individual chemical reactions. Each metastable ion represents an individually defined transition for which masses and abundances of both products and reactants can be specified. The analytical advantage that accrues is based on the greater information inherent in a chemical rather than a physical approach. An ion once formed is not simply measured to establish its mass and its relative abundance, but rather is processed in experiments designed to define its chemical reactivity.

Independent sequential analyses. In order to explain the basis of an MS/MS experiment consider the basic reaction sequence shown in reaction 1.

$$M_1^+ + N \ ----> \ m_2^+ + m_3 + N \hspace{3cm} (1)$$

The goal is to establish the masses of both the products and reactants of this chemical reaction. We thus require an analyzer to establish M_1^+, and a second analyzer to establish m_2^+. The mass of the neutral m_3 is defined by difference. The reaction takes place between the two analyzers, aided by energy added to the reactant ion in this region (vide infra). With the addition of a source (S) and the detector (D), and denoting the reaction as *, the block diagram of a simple MS/MS instrument is established (Figure 2). This diagram also explains the various experiments available in MS/MS. The salient points are: 1) there are two mass analyzers to characterize the reaction; 2) the transition from reactant to product occurs between the analyzers; 3) the analyzers operate independently.
    There must be a source of energy in order to initiate the reaction between the analyzers beyond the inherent metastable ion abundances. Typically the interanalyzer region is fitted with a collision cell which contains about a millitorr of target gas N (often nitrogen or helium). The incoming ion collides with the target gas, transforming some fraction of the kinetic energy of the ion into internal energy which then causes fragmentation.

Resolution. In MS/MS, each of the two independent mass analyzers can be operated at a low resolution while retaining a high overall selectivity. Since extraction of the highest resolution from a given analyzer requires a disproportionate effort, the solution of demanding analytical problems is simplified. By analogy, a chromatographer reduces the performance requirements of a single stage separation of a complex mixture by adding a simple sample prefractionation. The same general principle is apparent in MS/MS.

Signal-to noise. Mass spectrometers are extraordinarily sensitive devices, having the ability to analyze nanogram amounts of sample. MS/MS, as discussed above, deals with the chemistry of ionic reactions, and thus it is often chemical rather than electronic noise that establishes the limit of detection (6). Chemical noise is produced by matrix constituents other than the sample, the reactions of which may be insufficiently resolved from the sample reaction of interest. The role of the analytical chemist is to design the MS/MS experiment to provide the best possible discrimination against the chemical noise in the system. The problem is a significant one; in complex mixtures, the matrix constituents are present in great excess, and their reactions are unknown. However, the use of several stages of independent mass analysis can provide a very high signal-to-noise ratio. Table I summarizes the operation of a generic species of analyzers given an initial mixture with equal parts of signal and noise. For this example, each analyzer passes 50% of the signal but only 10% of the

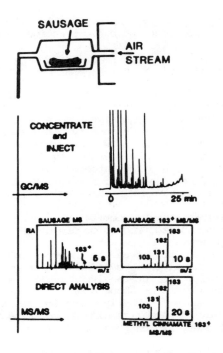

Figure 1. Comparison of the time scales of the procedures of GC/MS and MS/MS analysis of volatile flavor compounds emitted from sausage samples (19).

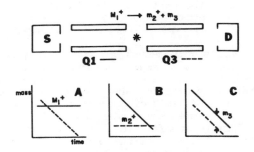

Figure 2. Simple diagram of an MS/MS instrument and three scanning modes based on changes in mass between the parent and the daughter ion. See text for details.

noise. Even with this crude differentiation, after two stages of
analysis, the signal to noise ratio has increased from unity to
25:1. With further analysis, it rises rapidly even as the total
signal level decreases. In MS/MS, the choice of ionization method,
ion polarity, and the mass analysis itself all contribute to the
differentiation of signal from noise, and the stepwise enhancement
is often several hundred to one rather than 5:1 as shown in this
example.

Table I. Signal to Noise Enhancements with Multiple Analyses

Number of stages	Intensities		Signal-to-noise ratio
	Signal	Noise	
0	1000	1000	1
1	500	100	5
2	250	10	25
3	125	1	125

Types of Experiments

The analysis in a mass spectrometer is not based on the mass of an
ion, but rather on its mass to charge ratio, m/z. Thus if either
the mass or the charge of the ion is altered in an MS/MS
experiment, the change can be followed by the second mass
analysis. The experiments of MS/MS can be subdivided into those
which involve changes in mass or changes in charge. There exists a
third category of experiments which involve changes in reactivity
independent of mass and charge, but these experiments will not be
discussed in this chapter.

Changes in mass. The most common experiments in MS/MS are based on
changes in mass. These are summarized in Figure 2. Assume a
complex mixture has been introduced into the source, and that ions
are formed corresponding to each constituent of the mixture. The
first mass analyzer selects ions of a specified mass which are
passed into the collision region between the analyzers. Here the
additional energy imparted by collision causes the breakup of this
parent ion into smaller fragment ions. The masses of the fragment
ions, termed daughter ions, is established by scanning the second
mass analyzer. The resulting spectrum is called a daughter ion
MS/MS spectrum, and consists of all of the fragment ions from a
selected parent ion (Figure 2a).
    Since the mass analyzers operate independently, it also is
possible to set the second mass analyzer to pass only daughter ions
of a selected mass to the detector. The first mass analyzer is
then scanned across the mass range. A signal at the detector is
noted when the first mass analyzer passes a parent ion that
fragments to the specified daughter ion. The spectrum that is
obtained is called a parent ion MS/MS spectrum, and consists of all
the precursor ions of a specified daughter ion (Figure 2b).
    If both mass analyzers are scanned at the same rate with a

constant mass offset x between them, then signals will be observed at the detector whenever a parent ion passing through the first mass analyzer produces a daughter ion with a mass x daltons less than the parent ion. The spectrum obtained is called the constant neutral loss MS/MS spectrum, and consists of all the parent ions in the parent ion/daughter ion pairs related by the loss of a neutral of specified mass (Figure 2c).

The three MS/MS experiments described above provide different information in complex mixture analysis. The daughter ion MS/MS spectrum is often obtained when targeted compound analysis is performed. The parent ion selected corresponds to the targeted component, and the daughter ion spectrum obtained from the mixture is compared to that obtained for the authentic targeted compound introduced to the source under the same conditions. In this way, the presence of the target can be established and often quantitated. Parent and constant neutral loss MS/MS spectra are more often used for identification of functional groups, and can be used for both targeted compound analysis, or for a completely unknown mixture. Experience often shows that there are characteristic daughter ions or neutral losses that occur for specific functional groups. For instance, 149^+ as a daughter ion is characteristic in the MS/MS spectra of phthalates. A scan for parent ions of the specified daughter ion 149^+ would thus pinpoint all of the various phthalates present in a mixture, regardless of whether each was known to be present or not. Similarly, the protonated molecular ions of carboxylic acids typically form daughter ions by loss of carbon dioxide. A constant neutral loss MS/MS spectrum with the offset specified as 44 daltons (the weight of the neutral fragment CO_2) will pinpoint parent ion/daughter ion pairs that undergo a chemical reaction typical of carboxylic acids, again without prior knowledge of their presence.

Changes in charge. The collisions used to add energy to ions and cause fragmentation also may cause changes in charge of the ion. Singly-charged ions can be oxidized to doubly-charged ions if the energy imparted by the collision is higher than the second ionization potential of the molecule (charge stripping, Equation 2). Doubly-charged ions passed into the collision cell can be reduced to the singly-charged ion (Equation 3) in a process known as charge exchange; the neutral involved in the collision acquires the balancing charge. Finally, negative ions can be converted into the corresponding positive ions in the oxidation process known as charge inversion (Equation 4).

$$M_1^+ + N \longrightarrow M_1^{2+} + e^- + N \qquad (2)$$

$$M_1^{2+} + N \longrightarrow M_1^+ + N^+ \qquad (3)$$

$$M^- + N \longrightarrow M^+ + 2e^- + N \qquad (4)$$

The energetic requirements for these reactions are different from those of the reactions which involve changes in mass. For the most part, these reactions are observed in high energy collisions. They have predominately been used for studies of ion structure, but have recently been used for complex mixture analysis. This expansion is based on the relationship between ion structure and the facility of charge changing reactions. Nitrogen-containing compounds, for instance, form doubly-charged ions more readily than other classes of compounds. An experiment based on the reaction of doubly charged ions can thus be specific for nitrogen-containing ions.

Analytical Characteristics of MS/MS

MS/MS was first considered as a replacement for GC/MS. Its true character as a complement to that method is now realized, and the most demanding of analytical problems often require the full differentiating power of a GC/MS/MS combination. The choice between GC/MS or MS/MS for a particular application must rest on relative merits of sensitivity, selectivity, and speed, each of which will now be briefly discussed.

Sensitivity. It is misleading to assign a single value of sensitivity for either GC/MS or MS/MS without specifying the problem, the instrument, the experiment, the spectrum, and the operator. In general, for modern instruments, analyses of compounds at the nanogram level can be considered routine for either technique. Lower limits of detection are available with special attention to the experiment. For example, in GC/MS, selected ion monitoring is used to increase sensitivity. In this experiment, the mass analyzer no longer scans across the full mass range, but rather integrates signal in a few mass windows corresponding to ions of interest. The effective resolution of the chromatographic separation is usually increased. The selected ion monitoring technique is useful when compounds of particular interest are known to produce characteristic ions in the source. Thus Harvey (7) demonstrated that the trimethylsilyl (TMS) derivatives of diphenylpropanoids form characteristic ions at m/z 266 (Scheme 1). Setting the mass analyzer of the spectrometer to pass only m/z 266 pinpoints the elution of such compounds from the gas chromatographic column. Sensitivity for selected ion monitoring experiments typically is reported in the low picogram level, although in favorable cases, low femtogram sensitivity can be achieved.

   In MS/MS, the selected ion monitoring experiment is transformed into selected reaction monitoring. Both mass analyzers are set to pass specified parent ion/daughter ion pairs. As in selected ion monitoring, there is a time advantage as the mass analyzers are not scanned but rather integrate signal. There is an added specificity over selected ion monitoring in that both the reactant and the product are specified. If sample introduction is via the direct probe, the vaporization profile provides a third parameter via which the compound can be identified. MS/MS has been used for the identification of targeted compounds in complex mixtures at the nanogram level (8). At lower levels, matrix constituents affect the precision of the response, necessitating either higher resolution measurements or sample cleanup. The latter route has been successfully pursued under the guise of GC/MS/MS down to levels of 10–100 pg in pharmaceutical applications (9).

Specificity. Yost (10) has compared the relative informing power of GC/MS and MS/MS. The informing power of an analytical procedure is expressed mathematically in terms of the number of "bits." A capillary gas chromatography column of 10^5 plates combined with a quadrupole mass spectrometer with unit mass resolution up to 1000 daltons provides an informing power of 6.6 x 10^6 bits. An MS/MS instrument comprised of two sequential quadrupole mass analyzers of the same performance provides an informing power of 1.2 x 10^7 bits. Within the limit of the assumptions made, the informing powers are identical. There are, however, various experimental parameters in MS/MS which can be used as additional resolution elements. These include energy resolved, pressure resolved, and angle resolved MS/MS experiments (4). These parameters are balanced by variation of gas chromatographic conditions, such as the stationary phase chosen, the temperature program followed, and additional steps of sample purification and pretreatment. In summary, both GC/MS and MS/MS are powerful enough to solve most analytical problems, both generating the extensive data sets characteristic of such combined methods (11).

Speed of Analysis. The introductory example focussed on the speed of MS/MS analysis of volatile compounds. However, there are several aspects of analytical speed of interest in MS/MS. The first, and that which has received the most attention, is the time required for analysis. The analogy between GC/MS and MS/MS involves the comparison of retention times through the gas chromatograph with ion flight times through the first mass analyzer. The former occupies between 10^1 and 10^3 s, and the latter on the order of microseconds. In the specific example of targeted compound analysis in a complex mixture, MS/MS can offer a significant time advantage in the examination of a large number of samples. Glish has shown that the analysis of mixtures using a preset protocol of selected reaction monitoring can occur at near the rate of sample introduction into the source of the instrument (12).
     The time advantage of MS/MS also is exemplified by the ability to select from a mixture of ions in the source any parent ion, in any order, and to return as necessary to that parent ion for precise measurements. This independence of access persists for the duration of the sample residence time in the source. This is in marked contrast to the situation in GC/MS, where each sample is available for examination by the mass spectrometer only during the retention time window. To repeat the measurement, the entire sample must be reinjected. The source residence time for most samples introduced via the direct insertion probe is on the order of a minute or two, depending on the temperature of the source and the rate of heating of the probe tip itself.
     The final aspect of speed to be considered is the information flux. In MS/MS, the sample may be available for minutes rather than the seconds corresponding to the width of a gas chromatographic peak. In the absence of a preset protocol, experimental decisions must be made in real time. For example, what parent ion should be selected for a daughter ion MS/MS spectrum? Does the spectrum of parent ions from the source change with probe temperature? What should the collision energy and

pressure be? The newest generation of data systems can make such decisions automatically within certain preset limits (13). The rate at which information must be obtained and processed in these latter MS/MS experiments is much higher than that in a GC/MS experiment.

Applications to Flavor Compounds

The use of gas chromatography/mass spectrometry in food and flavor analyses is now well-established, and reviews are plentiful (14-16). By contrast, the use of MS/MS in this area is less widespread. In part this has been due to the longer availability of commercial GC/MS instruments as opposed to MS/MS instruments, but also in no small part due to the enormous success of the GC/MS method itself. Food and flavor analyses deal perforce with the identification and quantitation of volatile components of complex mixtures. Gas chromatography, especially in capillary form, is able to separate such compounds with high efficiency. The addition of the mass spectrometer allows the identification of the eluted compound at the very low levels found in many food and flavor mixtures. As in GC/MS, the analyst using MS/MS must be concerned with sample handling (collection, treatment, and contamination) and sample analysis (ionization method and mass measurement).

Sample collection, treatment, and contamination. In GC/MS, sample treatment is often extensive. In preparation for GC/MS analysis of nutmeg, Harvey (7) ground 100 mg of nutmeg to a fine powder and extracted for an hour with ethyl acetate. The filtered extract was frozen to precipitate triglycerides, filtered again, and then derivatized overnight with a standard trimethylsilylation reaction. By contrast, in the MS/MS analysis of nutmeg by Davis (17), 10-50 mg of ground nutmeg are loaded into a glass capillary, introduced directly into the source of the mass spectrometer, and vaporized by a short heating program.

A constant concern in the analysis of flavor components is alteration and contamination of the sample. Losses of volatile components are a major problem. The extensive sample preparation involved in GC/MS offers ample opportunity for transformations and losses because of sample handling and exposure to chemical derivatizing reagents. In MS/MS, sample handling is often reduced and the chances for outside contamination minimized. Sample carryover, a problem during extraction procedures for GC/MS, is not eliminated in MS/MS, but evolves into a problem of source contamination. This problem was severe in some early MS/MS work, but now seems under control with the use of removable ion volumes in the source. The concentration and homogenization of sample that occurs in GC/MS pretreatment is not available in MS/MS. Sample inhomogeneities thus become of much greater concern. Sample to sample variation is already fairly high in samples of natural origin, and the extensive application of MS/MS may require more careful sampling procedures than currently employed. For trace analyses, a simple form of sample pretreatment is often employed in MS/MS to concentrate the sample and to preserve the cleanliness of the ionization source.

Sample ionization. Requirements for sample ionization are much
more severe in MS/MS than in GC/MS. For MS/MS, the ionization
method should create one ion for each component, and the structure
of the ion should be the same as that of the neutral surrogate.
Electron ionization usually does not fulfill these requirements,
since the ions formed often include those from rearrangement
reactions, and the degree of fragmentation is excessive. Chemical
ionization provides the requisite single ion for each component of
the matrix in the form of the quasimolecular ion $(M+H)^+$.
However, chemical ionization is sensitive to source parameters and
matrix effects, and these problems are exacerbated by the direct
introduction of a complex mixture into the source. The effects can
be compensated to some degree by the use of an isotopically
labelled internal standard for quantitative work. In the analysis
of unknowns in complex mixtures, the nature of the source chemistry
should be a constant concern.

Sample Derivatization. The derivatization of nutmeg constituents
described by Harvey (7) is designed to increase the volatility and
stability of the components so that they can be separated in the
gas chromatograph. With direct probe introduction, MS/MS is
usually able to deal with samples of lower volatility; hence,
derivatization is not required. Direct probe temperatures reach as
high as 400° C, vaporizing many samples directly into the vacuum
of the mass spectrometer source. Derivatization is used in MS/MS
for the somewhat different purpose of imparting a specific chemical
reactivity to the analyte.
    Consider the trimethylsilyl derivatives often used to increase
volatility and stability. The electron ionization mass spectra of
these derivatives often contain fragment ions such as the
trimethylsilyl cation itself TMS^+, or fragment ions due to losses
of neutral species containing the trimethylsilyl moiety. The same
fragmentation reactions are expected in the MS/MS spectra of these
derivatives. Treatment of a mixture with a silylating reagent
converts free hydroxyl groups to their -OTMS derivatives, and then
a second labelled silylating reagent converts amino groups to their
-NH-d$_9$TMS derivatives. A parent ion scan of TMS^+ (at m/z 73)
pinpoints all of the precursor ions that contained a free hydroxyl
group. A parent ion scan of d$_9$-TMS^+ pinpoints the precursor
ions with a reactive amino group. Common ions in the two MS/MS
spectra represent molecules that contain both reactive groups.
    A derivatization scheme involving a constant neutral loss
MS/MS scan has been described by Zakett (18). Phenols and amines
react with acetyl chloride to form acylated derivatives which
commonly lose the neutral fragment ketene in the MS/MS reaction. A
neutral loss scan for the loss of 42 daltons will thus indicate the
molecular weights of any compound which has undergone
derivatization. The strategy was used successfully in the analysis
of these functional groups in a synthetic fuel sample (18).
Applications to flavor compounds have not yet been reported, but
will undoubtedly be extensively exploited considering the diversity
of derivatization chemistry already developed.

Applications

Food aromas. Labows and Shushan (19) have reviewed the direct
analysis of food aromas by a commercial MS/MS system using an
atmospheric pressure ionization source. The sample inlet is a
simple all-glass device that collects volatile components emitted
by food materials and directs them into the mass spectrometer.
Losses due to sample preparation are minimized, as are absorption
or decomposition problems associated with chromatographic
fractionation. Profiles of aroma compounds obtained by this method
are claimed to be more accurate than those obtained using other
analytical methods. Because of the high discrimination against
chemical noise in the MS/MS system, detection limits can be very
low, reported as 0.5 ppb for ethyl butyrate, 0.8 ppb for linalool,
and 45 ppb for limonene. These limits were established with
daughter ion MS/MS spectra.

Figure 3 shows the correlation between the daughter ion MS/MS
spectrum of authentic nootkatone (with the ion at m/z 219 selected
as the parent ion) and the ion at the same mass emitted directly
from a grapefruit. The match between the two spectra confirm the
presence of this targeted compound in the emitted volatiles. It
also has been identified in the volatiles from intact oranges. The
experiment can be completed in less than a minute, without sample
preparation. Note that the unit mass resolution of the triple
quadrupole instrument allows an accurate assignment of abundances
for daughter ions of adjacent mass in this MS/MS spectrum. A close
examination of the spectra show that the match between the
authentic and the target compound is not perfect. Either the
instrumental parameters were not constant, or there is an
additional component at m/z 219 in the volatiles emitted by the
grapefruit. It is at this stage that a simple prefractionation
experiment, or an alternative ionization method becomes necessary
to establish the number of components present at this mass.

Other MS/MS experiments were used to give information useful
for functional group characterization. Fragmentation to m/z 18
(NH_4^+) is indicative of amines, and m/z 19 (H_3O^+) is a
typical fragment ion from alcohols. Thus a parent ion scan for
these daughter ions pinpoints compounds of these groups in the
emitted volatiles. Acetate esters produce daughter ions at m/z 43
and m/z 61. A parent ion scan for the latter produces the daughter
ion MS/MS spectrum shown in Figure 4, which is the sum of all the
parent ions of all of the acetate esters in the volatiles emitted
from a banana. The base peak at m/z 131 represents the parent ion
of isoamylacetate, known to have the characteristic banana odor.
Figure 5 is the neutral loss scan for loss of 44 daltons, a
familiar loss from negative ions of carboxylic acids. Thus the
ions at m/z 87, 89, and 121 are most likely from butanoic or
pyruvic acid, lactic acid, and benzoic acid, respectively. The
identity of these ions are confirmed by examining the daughter ion
spectra of the authentic compounds and the peaks obtained in the
direct analysis of the sample, in this case a Teewurst sausage.

Direct analyses of volatiles has been suggested as a means of
screening food products that might otherwise pass agricultural
borders. The sensitivity seems to be sufficiently high for this
purpose, and the MS/MS analysis possesses the requisite speed and

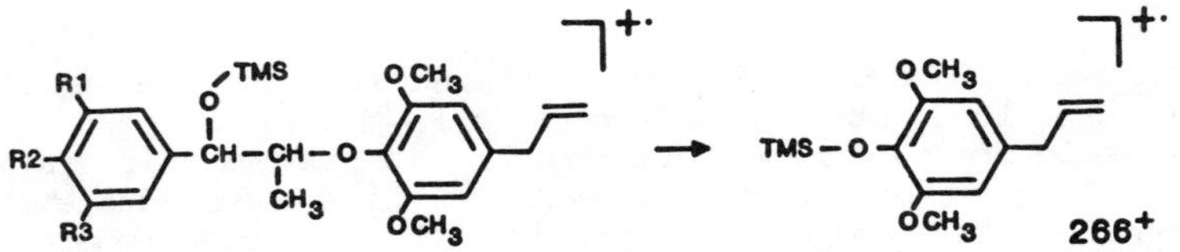

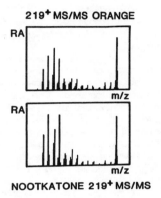

Scheme 1.

219+ MS/MS ORANGE

NOOTKATONE 219+ MS/MS

Figure 3. Daughter ion MS/MS spectra of suspected nootkatone emitted from grapefruit and the authentic compound (19).

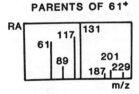

PARENTS OF 61+

Figure 4. Parent ion MS/MS spectrum for acetate esters emitted from a banana, representing an example of functional group screening by MS/MS (19).

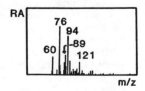

Figure 5. Neutral loss MS/MS spectrum for loss of 44 daltons (carbon dioxide) which pinpoints (M-H)⁻ molecular ions for carboxylic acids. Lactic acid and benzoic acid are identified at 89⁻ and 121⁻, respectively, although these are not the largest peaks in the spectrum (19).

specificity for real time analysis. A comprehensive study of the
interferences that might be expected in such use would be needed to
evaluate the suitability of this technique.

Spice analysis. Davis has studied the composition of nutmeg using
MS/MS (17). Nutmeg has been extensively studied because of the
large number of psychoactive species alleged to be present. This
study is noteworthy because of the use of both high energy and low
energy MS/MS to acquire daughter ion spectra for the various
compounds contained within the nutmeg, and the use of programmed
thermal desorption from the direct inlet probe of the mass
spectrometer in order to effect a crude distillation of the
sample. Isobutane was used as the reagent gas in order to ensure
that most of the constituents form protonated molecular ions
(M+H)$^+$ in the mass spectrum with a minimum of fragmentation.
Charge exchange was used as an alternative method of ionization in
order to form parent ions in an independent manner, and examine the
daughter ion spectra of the complementary parent ion M$^{+\cdot}$.

The isobutane chemical ionization mass spectrum of nutmeg
obtained at a probe temperature of 150°C differs from that
obtained at 200°C. Higher mass volatiles are not evaporated into
the source until the temperature of the probe is elevated to the
higher temperature. As with many analyses of this type, the amount
of sample is not a limiting factor. The temperature can thus be
held steady for several minutes at a given value, allowing several
independent MS/MS experiments to be completed. At higher probe
temperatures, thermal degradation of the sample can become a
problem.

Comparison of the daughter ion MS/MS spectra of authentic
4-allyl-2,6-dimethoxyphenol at m/z 195$^+$ and the same mass ion
from the nutmeg sample is presented in Figure 6. The spectra are
sufficiently similar that the presence of this compound in nutmeg
can be confirmed. Of particular value is the sharp charge
stripping peak at 97.5 on the mass scale. This is the product of
an oxidation reaction of the singly charged 195$^+$ to the
doubly-charged 195^{2+} as a result of the high energy collision.
This reaction occurs frequently with nitrogen- and
oxygen-containing compounds. Two points should be noted. First,
the match, although close, is not exact. This indicates that not
all of the ion current at this mass is due to this compound alone,
as in the case described above. Secondly, the width of the peaks
for the daughter ions are very wide, compromising both the
assignment of masses and the relative abundances. This is a
consequence of the instrument used, which was a reverse-geometry
sector instrument (20). Daughter ion analysis is accomplished with
a kinetic energy analyzer, which mirrors the kinetic energy release
observed as a consequence of the fragmentation reaction. Although
this value can be used as a probe of the mechanism of the
fragmentation itself, it is a disadvantage in MS/MS work for these
reasons.

It is known that nutmeg contains diphenylpropanoids of cyclic
and acyclic forms. The acyclic form fragments to characteristic
daughter ions at m/z 193, and the cyclic form to daughter ions at
m/z 203 (Scheme 2). These daughter ions can be set as products in
a parent ion scan to examine the entire nutmeg mixture for parent
ions of these classes of diphenylpropanoids. Figure 7 shows the

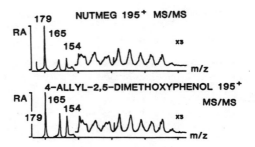

Figure 6.   Daughter ion MS/MS spectrum of the protonated molecular ion from 4-allyl-2,6-dimethoxyphenol as an authentic compared to the spectrum obtained from an ion of the same mass formed directly from a nutmeg sample (17).

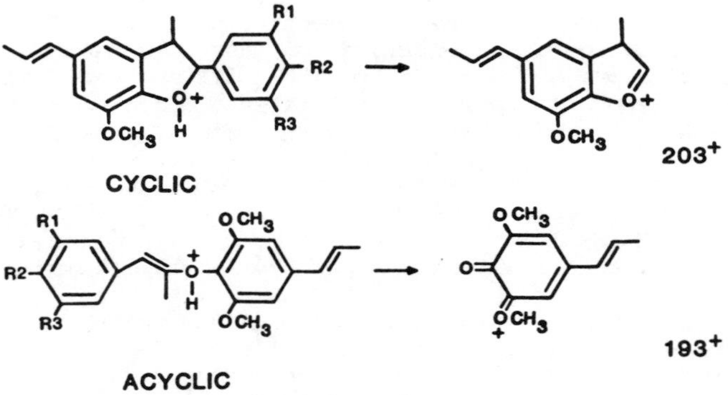

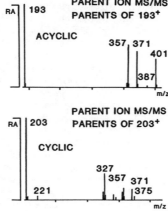

Scheme 2.

Figure 7.   Parent ion scans for two isomeric forms of diphenylpropanoids found in nutmeg. The common ions at m/z 355, 357, and 371 indicate the presence of both forms of the compound at those masses (17).

result of this experiment. Parent ions at 357^+, 371^+, 387^+, and 401^+ are indicated to be acyclic diphenylpropanoids. Cyclic forms are indicated in the parent ion scan at 327^+, 341^+, 355^+, 357^+, 371^+, and 375^+. The common parent ions 355^+, 357^+, and 371^+ are clearly indicated as consisting of both forms of diphenylpropanoid structures. The parent ion at 355^+ is thought to be a dehydrodiphenylpropanoid derivative of myristicin, identified for the first time in nutmeg.

Problems and Potentials of MS/MS

Problems. The simultaneous introduction of all the components of a mixture into the source of the mass spectrometer constitutes one of the strengths of MS/MS, but is also the cause of several problems. First is the problem of sample carryover and source cleanliness. One of the tenets of normal operation of a mass spectrometer is to introduce as little sample as possible, while in MS/MS, the sample sizes are often quite large -- up to several mg in trace analyses. Newer instruments are designed for easier source cleaning or utilze ion volumes which are replaceable through the direct probe inlet. It is possible to change the entire source volume with each sample and thus minimize this problem. Less easily ameliorated is the phenomenon known as the "matrix effect". Ideally, the chemical ionization source produces one ion for each constituent of the mixture, and the relative abundances of the ions formed are in proportion to the amount of constituent present. The "matrix effect" is a term that describes enhancement or suppression of ion signal for a single component due to the presence of the matrix. This is an insidious problem because the matrix is not characterized, and may change from sample to sample. For targeted compound analysis, the usual solution is to employ an isotopically labelled internal standard that is introduced into the mixture as a whole. Quantitation of the signal of interest is then derived from the relative abundances of the ions corresponding to the labelled and unlabelled forms of the analyte. Since the unlabelled/labelled ion pair persists in many of the daughter ions formed by collision, several confirming ratios can be obtained in a single experiment.

The matrix effect also may be evident in chemical noise which persists in the spectrum, and is far greater in analyses near the detection limit. This was highlighted in the paper by Bursey (21) in which a matrix effect in the determination of polychlorinated organic compounds was found. Direct probe MS/MS results were systematically high compared with those from GC/MS or GC/MS/MS. The rapid throughput possible in an MS/MS screening protocol is not obtained without concomitant risk.

MS/MS is an empirical method of analysis. As is evident from the examples presented, the interpretation of a daughter ion MS/MS spectrum is often based on the same correlation principles derived from electron and chemical ionization mass spectrometry. More often, the comparison of the spectrum obtained to that of the authentic compound is used for identification. This is a fundamentally unsatisfying procedure. While electron and chemical ionization spectra can be compared to a spectral library which has been compiled over the past thirty years, no comparable library of MS/MS spectra exists. Data systems may be used within individual laboratories to create libraries of spectral data which are then searched in the usual manner, but national and international databases are nonexistent. Several obstacles must be overcome to create such bases. First is the division of MS/MS spectra into those resulting from high energy collision processes (as on sector mass spectrometers) and those obtained under low energy collision conditions as prevail on the multiple quadrupole instruments. The spectra thus obtained are often similar, but perhaps not to the point where a cross correlation can be drawn. Charge stripping and charge inversion are processes that are confined largely to high energy MS/MS spectra, for example. The less than unit mass resolution of the reverse geometry sector instruments for daughter ion MS/MS spectra would be a problem in reducing spectra into a standard library form, as both the masses and the relative abundances of the daughter ions are known with a limited certainty. Finally, the low energy daughter ion MS/MS spectra are affected by instrument parameters such as the collision energy and the collision gas pressure. To date, no general standard of operation has been accepted. While these spectral effects are valuable in extracting additional information from the MS/MS spectrum, they do represent a significant obstacle to the standardization of MS/MS libraries. The most hopeful direction might be in advanced data systems with memory sufficient to accept all spectra obtained, and searching algorithms sophisticated enough to deal with multiple spectra of a single compound.

As mentioned earlier, quantitation with MS/MS is often carried out with internal standards. Without standards, the accuracy and precision of quantitation is reduced due to matrix effects. Rough estimates can be quickly obtained with MS/MS. For many problems, this information is more than sufficient. For instance, new drugs are often derived directly from plant material. One of the first questions asked is the relative concentration of the desired material in the various plant parts. MS/MS has been used to provide the approximate amounts of the targeted compound in roots, stems, petals, or flowers. The plant tissue with the highest concentration of compound is then extracted.

In food and flavor analyses, the accuracy of the quantitative data required from MS/MS may be limited by the variability of the sample itself. In analysis of a large number of samples, MS/MS provides a quick indication of the amounts of compounds of interest. If variability falls outside of a preset tolerance, then only those samples are flagged for more exhaustive workup and a more rigorous quantitative analysis. This ability to focus analytical resources on samples of interest is a valuable property of the MS/MS experiment.

Potential. The development of MS/MS for analyses of foods and flavors will follow the same growth curve as it has in other applications. At this point, only the first part of the growth curve is evident. Several industrial laboratories are beginning to use MS/MS on a routine basis, and commercial pressure will drive the expansion of the method. The speed of the MS/MS analysis is a strong initial advantage. In the long term, it is likely to be the flexibility of MS/MS analysis that will sustain its use in these areas, and justify the high initial cost of the instrument. It is only a matter of programming the experiment that differentiates the analysis of nutmeg from the analysis of acetate esters emitted as volatiles from fruit. All these protocols can be developed and then stored within the data system as standard methods of analysis, and then called up as needed.

The ability of MS/MS to search for classes of compounds in a mixture will be as valuable in food and flavor analyses as it is in other complex mixture analyses, such as the pharmaceutical or environmental fields. Flavors are complex mixtures, but often consist of groups of chemically similar compounds. It is precisely the identification of these groups for which parent ion and neutral loss MS/MS experiments are particularly adept. This is a characteristic that is patently not available with GC/MS, which has been the usual method of analyses of these mixtures.

Literature Cited

1. Kruger, T. L.; Litton, J. F.; Kondrat, R. W.; Cooks, R. G. Anal. Chem. 1976, 48, 2113.
2. Kondrat, R. W.; McClusky, G. A.; Cooks, R. G. Anal. Chem. 1978, 50, 1222.
3. Kruger, T. L.; Litton, J. F.; Cooks, R. G. Anal. Lett. 1976, 9, 533–542.
4. McLafferty, F. W., Ed., "Tandem Mass Spectrometry"; Wiley: New York, 1983.
5. Cooks, R. G., Beynon, J. H.; Caprioli, R. M.; Lester, G. R., "Metastable Ions"; Elsevier: Amsterdam, 1973.
6. Kondrat, R. W.; Cooks, R. G. Anal. Chem. 1978, 50, 1251A.
7. Harvey, D. J. J. Chrom. 1975, 110, 91–102.
8. Plattner, R. D.; Yates, S. G.; Porter, J. K. J. Agric. Food Chem. 1983, 31, 785–789.
9. Richter, W. J.; Blum, W.; Schlunegger, U. P.; Senn, M. In "Tandem Mass Spectrometry"; McLafferty, F. W., Ed.; Wiley: New York, 1983.
10. Fetterolf, D. D.; Yost, R. A. Int. J. Mass Spectrom. Ion Proc. 1984, 62, 33–50.
11. Hirschfeld, T. A. Anal. Chem. 1980, 52, 297A.
12. Glish, G. L.; Shaddock, V. M.; Harmon, K.; Cooks, R. G. Anal. Chem. 1980, 52, 165–167.
13. Kirby, H.; Sokolow, S.; Steiner, S. In "Tandem Mass Spectrometry"; McLafferty, F. W., Ed.; Wiley: New York, 1983.
14. Issenberg, P.; Kobayashi, A.; Mysliwy. T. J. J. Agric. Food Chem. 1969, 17, 1377–1386.
15. Horman, I. Gazz. Chim. Italiana 1984, 114, 297–303.
16. Horman, I. Biomed. Mass Spectrom. 1981, 8, 384.
17. Davis, D. V.; Cooks, R. G. J. Agric. Food Chem. 1982, 30 495–504.
18. Zakett, D,; Cooks, R. G. In "New Approaches in Coal Chemistry"; Blaustein, B. D.; Bockrath, B. C.; Friedman, S., Eds.; ACS SYMPOSIUM SERIES No. 169, American Chemical Society: Washington, D. C., 1981; pp. 267–288.
19. Labows, J. N.; Shushan, B. Amer. Lab. 1983, 15(3), 56–61.
20. Beynon, J. H.; Cooks, R. G.; Amy, J. W.; Baitinger, W. E.; Ridley, R. Y. Anal. Chem. 1973, 45, 1023A.
21. Voyksner, R. D.; Hass, R. D.; Sovocool, G. W.; Bursey, M. M. Anal. Chem. 1983, 55, 744–749.

RECEIVED June 24, 1985

Automated Analysis of Volatile Flavor Compounds

Robert G. Westendorf

Tekmar Company, Cincinnati, OH 45222-1856

Volatile organic compounds present in foods have
a significant impact on flavor quality. The
analysis of these compounds can be quite difficult, since the sample is often not amenable
to direct GC injection, and the volatiles may be
present in very low concentrations while still
being important to flavor. The technique of
dynamic headspace sampling was used for the isolation and concentration of volatiles prior to
analysis by gas chromatography. Samples, which
may be heated, are purged with an inert gas,
sweeping any volatile compounds present out of
the sample. The volatiles are trapped on Tenax,
which is then thermally desorbed and backflushed
to inject the sample into the GC. Using new
instrumentation, this method was fully automated.
Samples run include fruits, fruit products, edible
oils, and oil-based foods. Detection limits in the
low part-per-billion range were obtained with 2–8%
reproducibility.

The instrumental analysis of flavor in a food material can be an
extremely difficult task. There are many factors that influence
flavor. Of these, one of the more important, yet also most
difficult to analyze, is the profile of volatile organic
compounds present. The difficulty arises from the fact that
there may be many volatiles present at very low concentrations
in a complicated matrix. A variety of methods have historically
been used for this analysis. The majority of these methods have
utilized gas chromatography (GC), differing in chromatographic
parameters and sample preparation techniques. Chromatographic
systems have evolved tremendously in recent years. Column
technology has advanced to a very high level of separation power.
Sample preparation techniques, on the other hand, have not
evolved as rapidly as GC technology. A number of different
preparation techniques have been used for the analysis of flavor

0097–6156/85/0289–0138$06.00/0

volatiles, including solvent extraction, steam distillation,
equilibrium headspace sampling, and dynamic headspace sampling.
Of these methods, dynamic headspace sampling is probably the
least well known, yet it has a number of advantages over other
techniques in use.

Methods utilizing dynamic headspace techniques were reported
as early as 1960 (1). These methods generally utilized cryogenic
traps, or cryogenically cooled traps containing column packing
materials (2) or molecular sieves (3). Dynamic headspace
analysis (DHA), also know as purge and trap analysis, utilizing
new porous polymers as trapping agents and incorporating
multi-port valves for flow switching was introduced for the
analysis of organic contaminants in water in 1974 (4). Various
forms of DHA have been used for a variety of food samples.
Fruits and juices have been investigated by a number of
researchers. Schamp and Dirinck (5) found over forty compounds in
a study of strawberry varieties, as well as finding 22 compounds
in Golden Delicious apples. Keenaghan and Meyers (6) reported on
the GC/MS identification of over twenty different compounds in
several varieties of apples and apple products. Additional work
with apples has been reported by Westendorf (7). This same paper
reported the analyses of dairy products, vegetable oils, and
artificial flavors. Edible oils have been extensively studied
since the presence of volatiles was first recognized as an
indicator of oil quality (8). Considerable work with oils and
oil-based foods using manual procedures has been reported by
Jackson et al. (9,10), Dupuy et al. (11,12), and Selke (13). In
1983 Roberts (14) first reported the use of an automated DHA
procedure for oil volatiles.

Principle of Operation

DHA is based on the partitioning of volatile compounds between a
sample and the vapor phase above the sample at a rate dependent
on a variety of factors. These include the volatility of the
subject compound, its solubility in the sample matrix,
homogeneity of the matrix, temperature, and sample container
configuration. In equilibrium headspace analysis, the sample is
sealed in a closed vessel and the volatiles are allowed to
equilibrate between the sample and vapor phase. An aliquot of
the vapor phase is then injected into a GC for analysis. In DHA,
the sample is purged with an inert gas, sweeping the volatiles
out of the sample container. The purge gas is then passed
through a short column containing a porous polymer adsorbent
which selectively retains the sample compounds while allowing the
purge gas and any water vapor to pass through. By purging in
this manner, the entire organic contents of the vapor phase can
be subjected to GC analysis, not just an aliquot. In addition,
since the purge gas is continually being removed from the sample
the concentration of organics in the vapor above the sample
remains essentially zero. This significantly enhances recovery
by promoting further partitioning of the volatiles into the vapor
state. After the purge step is completed, the adsorbent column
is heated to release the organics and backflushed via a 6-port
valve to sweep the sample to the GC. When using capillary

columns, a cryogenic focusing step is employed to sharpen the injection profile (15). Separation and detection are carried out in the GC, normally under temperature programmed conditions. A diagram of the flow scheme is illustrated in Figure 1.

Goals. The primary goal of this work was to adapt new instrumentation to the fully automated analysis of volatile flavor compounds in foods without compromising any other aspect of analytical capability. The goals of this work, in addition to automation, included:
1. Recover the maximum range of compounds possible, from very volatile gases to higher boiling, less volatile compounds.
2. Achieve the maximum sensitivity possible, preferably to part-per-billion (ppb) levels, since many compounds have important organoleptic qualities at low levels.
3. Achieve the best reproducibility possible.
4. Keep artifact formation to a minimum, to ensure that the volatile compounds found are those actually present in the sample.
5. Eliminate cross-contamination between samples.

Achievement of goals 1-5 was considered necessary for the primary goal, total automation, to be of any practical value.

Instrumentation for Automated Analyses. All samples were run using commercially available automated DHA equipment interfaced to a microprocessor GC. The DHA apparatus consists of three parts: the basic concentrator (TEKMAR Model 4000), incorporating the purge system, switching valve, and adsorbent; a ten-position automatic sampler (TEKMAR Model 4200); and a capillary column cryogenic trap (TEKMAR Model 1000). Under normal conditions the operator first sets all operating conditions, loads the samples, raises the sample heaters (high performance electric mantles), and places the instruments in automatic mode. The automatic sampler advances to the first position and signals the concentrator to start. The prepurge, preheat, and purge steps are performed by the concentrator, which then sends a ready signal to the capillary cryotrap. When the cryotrap also receives a ready signal from the GC, it will cool to a preset temperature and signal the concentrator to start desorbing the sample. When desorption is complete the cryotrap will heatup to inject the sample and simultaneously output signals to start the temperature program of the GC and to start data acquisition on an integrator. The integrator will also read the sample position number from the automatic sampler via a BCD interface. While the GC run continues the concentrator will recondition and then cool the adsorbent trap. When the trap has cooled the automatic sampler will again advance and the above cycle will repeat.

Experimental

Materials. Samples were obtained from a variety of sources. Fruits and juices were purchased locally. All oils, peanut butter, and food starch samples were supplied by food processors. Flavor compounds used in the preparation of standards were

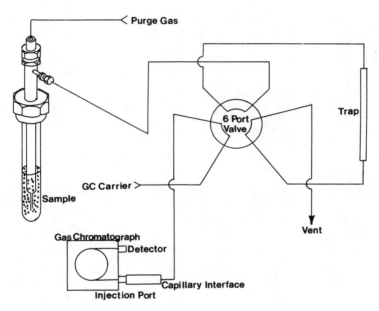

Figure 1:   Gas flow scheme of dynamic headspace gas
            chromatography system.

purchased from Aldrich (Milwaukee, WI) and used without further
purification.

Concentrator Conditions. The concentrator system used was a
TEKMAR Model 4000 with a TEKMAR Model 4200 ten position automatic
sampler. A TEKMAR Model 1000 Capillary Interface was used for
interface to the GC. The operating conditions were as follows:
sample sizes – peanuts, fruits: 1.00g, peanut butter, food
starch: 100mg, oils: 0.5ml, juices: 5.0ml. Sample temperatures
– fruits, juices, food starch: ambient (23°C), peanuts, peanut
butter: 100°C, oils: 150°C. Prepurge 3.5 min. at 50ml/min.
nitrogen, preheat 3 min. for 100° samples, 5 min. for 150°
samples, purge 10 min. at 50ml/min. nitrogen. The trap was
12" X 1/8" stainless steel packed with 24cm (approx. 150mg)
Tenax TA (Chrompack, Bridgewater, NJ). The trap was preheated
at 175°C and then desorbed at 180°C for 4 min. The trap was
baked at 225°C for 10 min. between runs. All valve lines and
sampler mounts were continuously heated to 100°C. The
Capillary Interface was cooled with liquid nitrogen for
cryotrapping the desorbed sample and heated for 10 seconds
for GC injection.

Gas Chromatograph Conditions. The GC was a VARIAN 6000 with a
flame ionization detector. The injection port was heated at
200°, the detector at 250°. The detector range was 10-12 AFS,
attenuation 64 except as noted. The column was 25m X 0.32mm fused
silica DB5 with a 1.0 micron film thickness (bonded SE54, J&W
Scientific, Rancho Cordova, CA). The carrier gas was hydrogen at
47 cm/s. The column was temperature programmed from an initial
temperature of 35°C, held for 4 min., to a final temperature of
200°C at 4°C/min.

Results and Discussion

Volatility Range. The range of flavor compounds recovered
depends on a variety of factors. The two primary factors are the
sample matrix and temperature. The matrix can affect recovery in
two ways. The first is the solubility of the flavor compounds in
the matrix. Compounds that have a poor solubility will be purged
more efficiently than compounds of high solubility. The second
is that volatiles may be physically bound in the sample. Flavor
compounds may be present primarily in the interior of a sample,
as is the case with coffee beans. In this case the mass transfer
rate of the volatiles through the matrix becomes a limiting
factor. For samples of this type, grinding or otherwise
homogenizing the sample can significantly increase the recovery.
However, care must be taken to avoid the possible loss or
adulteration of volatiles during this process. A cryogenic
grinding procedure developed for plastics (16) has been
successfully applied to food materials in the author's
laboratory.
        Increasing the temperature of a sample serves to increase
recoveries by increasing the vapor pressure of the volatile
compounds. The effect of increasing temperature on a corn oil
sample is illustrated in Figure 2. Temperature is generally the

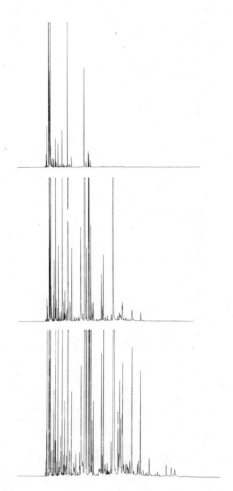

Figure 2:   Recovery of corn oil volatiles with increasing
temperature.

primary variable used in optimizing recovery. The recoveries of
a number of representative compounds from corn oil are listed in
Table I.

Table I. Representative Compounds in Corn Oil, 150°

Retention Time	Compound	Recovery	Reproducibility
1.03 min.	Butane	84%	10.5% RSD
1.36	Pentane	82	2.9
2.12	Hexane	74	5.3
3.67	Heptane	65	3.6
3.75	1-Butanal	47	3.6
5.33	1-Pentanal	50	4.9
6.16	Octane	61	4.0
6.24	1-Hexanal	51	4.9
6.58	2-Octene(a)	ND	7.9
7.67	trans-2-Hexenal	59	7.8
9.07	Nonane	55	3.4
9.15	1-Heptanal	43	3.4
12.06	Decane	50	1.9
14.79	Undecane	41	6.0
14.94	Amyl Alcohol(a)	ND	4.9
16.79	1-Hexanol(a)	ND	8.3
17.68	1-Nonanol(a)	ND	2.8

ND: not determined   a: tentative identification

A potential problem with heating some samples is the presence of
large amounts of water vapor. The Tenax adsorbent used is
hydrophobic, and does not trap water. However, water vapor will
condense on a cold Tenax trap. Excess amounts of water can
interfere with an analysis in two ways. If the amount of water
present is extremely high, it will condense on the Tenax in
sufficent quantity to physically block a significant portion of
the available trapping surface. This leads to reduced trapping
efficiency, degrading both sensitivity and reproducibility. For
samples of this type, the sample temperature must not be raised
above 65° to 95°, depending on the water content of the sample.
At 60°, however, even aqueous samples can be run without losing
trapping efficiency (17). A second way in which water may
interfere is in the GC separation and detection. Water may
degrade the column used, or interfere in the detection process.
Samples that have introduced sufficient water into the column to
extinquish the flame ionization detector have been encountered in
the author's laboratory. Since water is not trapped on the
Tenax, it is possible to remove most of it by passing dry
nitrogen through the trap before the desorption step (18). The
volatility range of the compounds that can be analyzed by dynamic
headspace concentration extends from organic compounds that are
gases at room temperature up to compounds containing about ten to
thirteen carbon atoms per molecule, depending on the number and
nature of any side chains. Table I lists a number of
representative compounds in a corn oil. Table II lists some

additional compounds commonly encountered in vegetable oil samples. Compounds typically found in fruit samples include primarily ethyl esters with a number of aldehydes and alcohols in the three to nine carbon range. Table III lists the GC/MS identifications of the compounds encountered in an apple sample (from ref. 6).

Table II: Additional Compounds Found in Vegetable Oils

Benzaldehyde	Benzyl Alcohol	1-Butenal
cis-2-trans-4-Decadienal	trans-2-trans-4-Decadienal	2-Decanone
2-Decenal	2,4-Heptadienal	Diacetyl
2-Heptanone	Methyl Ethyl Ketone	1-Hexanol
1-Nonanal	2-Nonanone	Octadiene
Octenal	1-Pentanol	

Table III: GC/MC Identification of Apple Volatiles (from ref. 6)

Compound	Compound
Butanal	N-Propyl Butyrate
Ethyl Acetate	N-Butyl Propionate
1-Butanol	N-Amyl Acetate
N-Propyl Acetate	Methyl Caproate
Methyl Butanoate	Ethyl-2-Methyl Butyrate
2-Methyl Butanol	N-Butyl-N-Butyrate
N-Hexanal	Ethyl-N-Caproate
Ethyl-N-Butanoate	N-Hexyl Acetate
N-Butyl Acetate	Isopropyl Hexanoate
2-Hexanal	1-Hexanol
N-Hexyl-N-Butyrate	2-Methyl Butyl Acetate
2-Methyl-2-Methyl-Propyl Butyrate	

Sensitivity. The sensitivity obtainable depends primarily on the efficiency with which a compound is recovered from the sample. As with the volatility range, recovery generally is most effectively increased by raising the sample temperature. Additional factors affecting sensitivity include trapping and desorption efficiencies, column resolution, interferences, and detector sensitivity. For oils the lower limit of detection for the majority of the compounds listed in Tables I and II is on the order of 1 to 100 ppb. For oil samples, nonane, which is often added as an internal standard, is detectable to less than 5ppb.

Reproducibility. Reproducibility is a third factor that depends primarily on recovery. As a general rule, reproducibility improves as recovery increases. For the majority of compounds for which the recovery is greater than 40%, the reproducibility will be on the order of 2 - 8% relative standard deviation (RSD). This number is strongly affected by column resolution, however, since many foods samples tend to give complicated chromatograms. As the recovery drops below 40% the reproducibility rapidly deteriorates. For very volatile compounds, the purge efficiency

is generally high, but trapping efficiency may become a factor.
The reproducibility obtained for pentane in corn oil is 2.9% RSD,
but the reproducibility for butane is 10.5% RSD. This suggests
that butane is not quantitatively trapped by Tenax. A number of
researchers are currently evaluating new sorbent materials, which
may be used alone or in combination with Tenax to improve the
analysis of compounds with lower molecular weights.
    When heating samples, an additional factor concerning
reproducibility is introduced. Not only must the heater control
be precise, but a time must also be allowed for the sample
temperature to equilibrate with the heater before starting to
purge. Samples placed in a heater at 150°C do not instantly
reach and equilibrate at 150°C. The rate at which the actual
sample temperature rises is extremely difficult to reproduce. If
purging is begun before the sample has equilibrated, the
resulting reproducibility may be acceptable, but can be improved.
By allowing a preheat time sufficient for the sample temperature
to equilibrate before purging, any temperature variances
resulting from heater variations, differing samples or sample
sizes (i.e. differences in heat capacity), or geometric
variations in the sample holder (e.g. solid chunks will have
different amounts of surface area contacting the walls of the
vessel) can be minimized.

Artifact Formation. The potential for the formation of artifacts
is present in virtually every analytical method. An advantage of
dynamic headspace analysis is that no solvent is used,
eliminating the greatest single source of artifacts.
Unfortunately, however, other mechanisms of artifact formation do
exist. The instrumentation must be designed to minimize any
possible artifacts. When running heated samples there are a
number of possible mechanisms for the formation of artifacts. For
compounds that are thermally labile the only method of preventing
their destruction is to maintain the sample temperature below any
possible breakdown point. For samples subject to oxidation, an
additional step is needed. The volatiles present in oils, for
instance, are primarily formed through oxidation reactions. At
the temperatures normally used to run oils, the sample will
rapidly react with any traces of oxygen present to form new
compounds or increase the concentrations of oxidation products
already present. This can be aggravated by the preheating step
used to achieve temperature reproducibility, allowing a greater
time for the hot sample to react with any oxygen dissolved in the
sample or present in the sample vessel. To remove the oxygen a
"prepurge" step is added in which the sample is purged for a
short time before heat is applied. Any oxygen present will be
removed from the sample and replaced by nitrogen. Note that this
prepurge gas must also be passed through the Tenax to trap any
compounds that may be purged at the low temperature. After
prepurging, the sample heater is turned on and the temperature
allowed to rise. The effects of the prepurging step can be noted
in Figure 3, which illustrates a corn oil sample run twice under
identical conditions with the exception of the prepurge step.
For further consideration of introducing artifacts, the
possibility of cross contamination between samples must first be
examined. It is possible in any instrumental method for part of

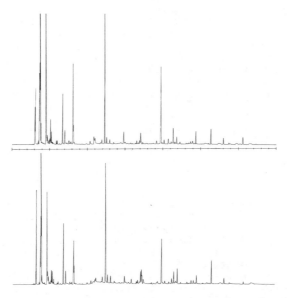

Figure 3:   Variances in corn oil sample without (top) and with (bottom) prepurge.

one sample to "carryover" into the next run, influencing the
results of the second run. With a concentrator system, the
possible causes of carryover are threefold: incomplete
desorption, condensation, and adsorption. Any possible traces of
sample components remaining on the Tenax trap as a result of
incomplete desorption are generally removed via a bake step, in
which the trap is heated to a temperature above that used for
desorption while passing gas through the trap. Condensation and
adsorption can occur in the valving and transfer tubing used.
These can be eliminated by heating the valve and lines. However,
this introduces the possibility of artifact formation via
catalytic decomposition on the hot surfaces in the instrument.
The effect of temperature differences is illustrated in Figure 4
on a soy oil sample. There was no measurable change for any
sample run between ambient and 100°C temperatures. However,
above 100°C some reactions began to occur. For the majority of
samples, 100°C was sufficient to prevent any carryover. For
samples containing large amounts of volatile components that
might have carryover at 100°C, higher temperatures can be used
for rapid cleanup.

Applications. In addition to oils and apples, a number of other
samples were used to evaluate the method. These included a
comparison of an orange to a food starch with added orange oil
(Figure 5). As is typical for many artificial flavors or
flavor-added samples, the primary differences occur in the early,
most volatile, portion of the chromatogram. Fruit samples can
also be evaluated for varietal differences (Figure 6), or for
evaluation of seasonal variations, ripening studies, or detection
of storage abuse. The evolution of flavor volatiles through two
different processes for peanut products is illustrated in Figure
7. The two chromatograms represent products made from the same
lot of raw peanuts, which had virtually no volatiles present
prior to processing. There are a number of additional sample types
for which DHA is currently being used. These include dairy
products, such as milk and cheese, carbonated beverages, powdered
drink mixes, beer, wine, coffee, meats, spices, grains, cereals,
and various forms of candy. Studies underway include evaluation
of processing techniques, shelf-life, packaging, and quality
control of products and incoming materials.

Conclusion

A method for the automated analysis of volatile flavor compounds
in foods is described. Volatile compounds are removed from the
sample and concentrated via the dynamic headspace technique, with
subsequent separation and detection by capillary column gas
chromatography. With this method, detection limits of low ppb
levels are obtainable with good reproducibility. This method has
experienced rapid growth in recent years. and is now in routine
use in a number of laboratories.

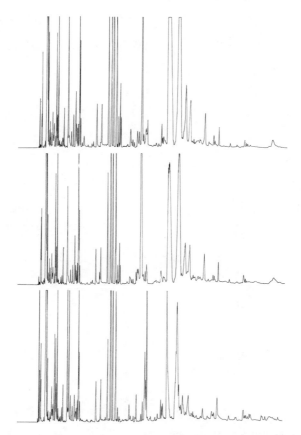

Figure 4: Soy oil sample run at different line temperatures.

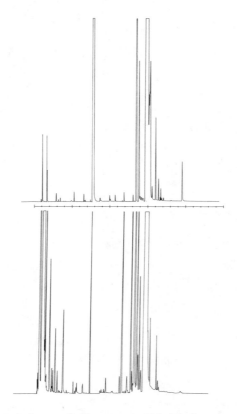

Figure 5: Comparison of flavor profiles of orange (bottom) and
orange-flavored food starch (top).

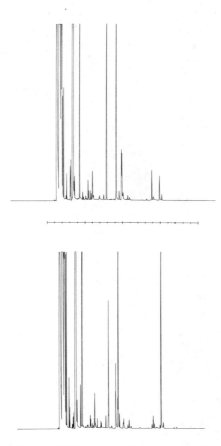

Figure 6: Varietal differences between white (top) and red
(bottom) grape juices.

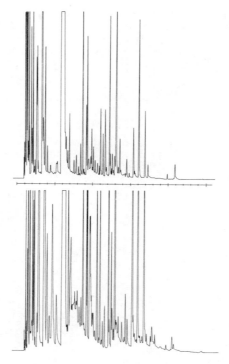

Figure 7:   Comparison of roasted peanuts (top) and peanut
butter (bottom) made from same lot of raw peanuts.

Literature Cited

1. Nawar, W.W., and Fagerson, I.S.; Anal. Chem., 1960, 32, 1534
2. Hornstein, I., and Crowe, P.F.; Anal. Chem., 1962, 34, 1354
3. Morgan, M.E., and Day, E.A.; J. Dairy Sci., 1965, 48, 1382
4. Bellar, T.A.; Lichtenberg, J.J.; Jour.AWWA 1974,66,739.
5. Schamp, N.; Dirinck, P.; in "Chemistry of Foods and Beverages: Recent Developments"; Charalambous, G.; Inglett, G., Eds.; Academic; New York, 1982; p25.
6. Keenaghan, J.; Meyers, M.C.; "Analysis of Volatile Organics in Foods and Beverages by Headspace Concentration-GC/MS"; presented at the 34th Pittsburgh Conference on Analytical Chemistry and Applied Spectroscopy; Atlantic City, N.J.; March, 1984.
7. Westendorf, R.G.; "Trace Analysis of Volatile Organic Compounds in Foods by Dynamic Headspace Gas Chromatography"; presented at the 35th Pittsburgh Conference on Analytical Chemistry and Applied Spectroscopy; Atlantic City, N.J.; March, 1984.
8. Scholz, R.G.; Ptak, L.R.; J.Am.Oil Chem. Soc. 1966,43,596.
9. Jackson, H.W.; Giacherio, D.J.; J.Am.Oil Chem. Soc. 1977,54,458-460.
10. Jackson, H.W.; J.Am.Oil Chem. Soc. 1981,58,227,231.
11. Dupuy, H.P.; Fore, S.P.; Goldblatt, L.; J.Am.Oil Chem. Soc. 1971,48,876.
12. Dupuy, H.P.; Fore, S.P.; Goldblatt, L.; J.Am.Oil Chem. Soc. 1973,50,340.
13. Selke, E.; J. Am.Oil Chem. Soc. 1970,47,393
14. Roberts, J.; "Semiautomated Dynamic Headspace Analysis of Vegetable Oil Volatiles"; presented at the 74th Meeting of the American Oil Chemists Society; Chicago, IL; May, 1983.
15. Capillary Column Use in Purge and Trap Gas Chromatography II, Use of the Model 1000 Capillary Interface"; Application Note B021684; Tekmar Company, Cincinnati, OH.
16. "Plastic Sample Preparation"; Application Note B081882; Tekmar Company, Cincinnati, OH.
17. Westendorf, R.G.; "Optimization of Parameters for Purge and Trap Gas Chromatography"; presented at the 32nd Pittsburgh Conference on Analytical Chemistry and applied Spectroscopy; Atlantic City, NJ; March, 1981.
18. "Purge and Trap Analysis Using a Photoionization Detector: Removal of Water Interference"; Application Note B042281; Tekmar Company, Cincinnati, OH.

RECEIVED June 24, 1985

Supercritical Fluid Extraction in Flavor Applications

Val J. Krukonis

Phasex Corporation, Nashua, NH 03060

Supercritical fluids are receiving increasing
attention as extraction solvents because of their
pressure-dependent solvent properties, often
displaying the ability to extract selectively one (or
a few) component(s) from a mixture; the chemical,
pharmaceutical, polymer, and food industries are
engaged in developing processes using these solvent
powers. Because selective, "dial-in-a-dissolving-
power" action can often be conferred to supercritical
fluids by virtue of the pressure applied on it, they
are attractive for extracting flavors present in
natural materials. Supercritical fluids, especially
carbon dioxide which is an active solvent at room
temperature, exhibit a frequent ability to
fractionate flavor and aroma components present as
complex mixtures of compounds in such materials as
pepper, ginger, allspice, and other spices. An
overview of research on supercritical fluid
solubility phenomena and on process operation
provides the framework to understand the extraction
capabilities of supercritical fluids; three examples
of ginger, allspice, and apple essence extractions
illustrate these capabilities.

By now, nearly every chemist has had some introduction to the
subject of supercritical extraction in one form or another, and it
would seem that after scores of papers, newsreleases, and trade
journal articles, only so much can be said about the background and
early findings, the thermodynamic interactions between dissolved
solutes and high pressure gases, the equations of state that can
correlate and predict solubility behavior, the many applications of
the technology (some of which are in flavors), the full scale
coffee and hops extraction plants now in operation, etc. What,
then, can a paper entitled "Supercritical Fluids - Overview and
Specific Examples in Flavors Applications" give that's new? -
hopefully, a different development of the historical perspective

and overview of supercritical fluid phenomena, and some recently
obtained data on flavor extraction and separation.

Some of the historical perspective is extracted (no pun
intended) from a previous paper of the author (1) and is expanded
with a chronological development of solubility phenomena based upon
an additional compilation of recent work on naphthalene-
supercritical solvent systems. The new data on flavor extraction
and fractionation point out the most unique feature of supercritical fluid solvents, viz., their often-demonstrated selective
dissolving power properties, a selectivity that is achieved because
the dissolving power of supercritical fluids is pressure-dependent
and can, therefore, be adjusted.

Historical Perspective

The ability of a supercritical fluid to dissolve low vapor pressure
materials was first reported by Hannay and Hogarth in 1879 (2).
They described their solubility experiments carried out in high
pressure glass cells, and they observed that several inorganic
salts (e.g., cobalt chloride, potassium iodide, potassium bromide,
ferric chloride) could be dissolved or precipitated solely by
changes in pressure on ethanol above its critical point ($T_c =$
234°C). For example, increasing the pressure on the system caused
the solutes to dissolve, and decreasing the pressure caused the
dissolved materials to nucleate and precipitate, in the words of
the authors, "as a snow."

In an historically interesting aside, there was initial
controversy about this finding after it was reported at a meeting
of the Royal Society (London). Professor W. Ramsay, in a
subsequent paper delivered to the Royal Society (3) stated that
based upon his reproduction of one of Hannay and Hogarth's
experiments, he concluded that "the gentlemen have observed nothing
unusual, but merely the ordinary phenomenon of solubility of a
solid in a hot liquid." Ramsay in the same paper also took to task
Dr. T. Andrews (the Andrews of carbon dioxide critical point
phenomena fame) for "purposely abstaining from speculating on the
nature of matter at the critical point, whether it be liquid or
gaseous, or in an intermediate condition, to which no name can be
given." Ramsay went on to describe some other of his own
experiments on critical point phenomena, and he concluded by saying
"I am inclined to think that carbonic anhydride, examined by Dr.
Andrews, is abnormal in this respect, but of this I am by no means
certain." As is now well known, and as also must have been known
to Andrews, carbon dioxide, the anhydride to which Ramsay refers,
is not abnormal in its critical behavior.

In still later presentations to the Royal Society (4,5),
Hannay responded to Ramsay's charge, asking "permission to point
out some errors into which Prof. Ramsay had fallen"; he did point
out those errors, and he presented results of more experiments on
dissolving solutes in supercritical fluids which for posterity
substantiated that the finding of the pressure-dependent dissolving
power of a supercritical fluid was indeed a new phenomenon. As
might be concluded from the excerpted statements, references 2-5
provide interesting reading of developments during a period of time

when critical point phenomena were still incompletely understood or accepted.

The pressure dependence of the dissolving power of a supercritical fluid is not limited to inorganic salt solutes, but is a relatively general phenomenon exhibited by all solid and many liquid solutes (as long as the solute is not infinitely miscible with the solvent). After the first report by Hannay and Hogarth, a number of other papers reported on solubility phenomena with a variety of supercritical fluid solvents and organic solid and liquid solutes. Solvents included carbon dioxide, nitrous oxide, the light hydrocarbons, and solutes covered the gamut of organic compounds, viz., aliphatics, aromatics, halogenated hydrocarbons, heteromolecules, triglycerides, and the like. Booth and Bidwell in 1949 presented an excellent review of nearly seventy-years of early research (6); several other comprehensive reviews cover later periods up to 1984 (7,8,9).

Supercritical fluid solvents received greatly increasing attention in the mid-to-late 70's, and papers described the process development effort on activated carbon regeneration (10), alcohol-water separation (11), chemotherapeutic drugs, flavor (12), and aroma extraction (13), and on many other separations. Many of these papers introduced the phenomenon of the pressure-dependent dissolving power of a supercritical fluid using naphthalene solubility data as an example, since naphthalene solubility has been studied more thoroughly than any other organic (or inorganic) compound. There is a substantial body of quantitative information available, and even more importantly, naphthalene's solubility behavior is representative of the behavior of many other compounds in supercritical solvents. Naphthalene solubility models the solubility of, for example, terpenes which have not been studied to as great an extent, but which are of commercial interest, or the behavior of triglycerides extracted from seeds with supercritical carbon dioxide, or the solubility of still more complex materials for which only scant data exist.

In 1948, a study of the solubility and phase behavior of naphthalene dissolved in supercritical ethylene was reported by two workers from The Netherlands, Diepen and Scheffer (14), and this now-classic paper was followed by two others from the same authors (15,16) and from others who reproduced the naphthalene solubility data of Diepen and Scheffer and extended the studies to other supercritical solvents. For example, Tsekhanskaya et al measured solubilities of naphthalene in ethylene and in carbon dioxide in 1962 (17,18). Others are King and Robertson in 1962, who studied the solubility of naphthalene in hydrogen and argon (19); Najour and King in 1970, naphthalene in supercritical methane, ethylene, and carbon dioxide (20); McHugh and Paulaitis, 1980, naphthalene in carbon dioxide (21), Kurnik and Reid, 1982, naphthalene in carbon dioxide (22); Schmitt and Reid, 1984, naphthalene in ethane, trifluoromethane, and chlorotrifluoromethane (23); and finally, Krukonis et al (24), and McHugh et al (25), who studied the solubility and phase behavior of naphthalene in the solvent supercritical xenon in 1984. Thus, as might be concluded from the long list of references cited, naphthalene solubility has been tested thoroughly, and a large amount of data exists on its

behavior over a wide range of pressure–temperature–gas conditions.
Because of this large data base, supercritical fluid applications
and process operation are quite frequently described in terms of
naphthalene solubility, the concept being extended to other systems
"by association"; some of the "association" will be developed
subsequently in this section.

Figure 1 is a graph of naphthalene solubility in carbon
dioxide at 45°C (15°C above the critical temperature of carbon
dioxide) taken from Reference 17. As is obvious from an
examination of the data, the solubility of naphthalene increases
dramatically when the pressure is increased beyond the critical
pressure of 73 atm. The solubility (given in units of grams/liter
in the reference) approaches about 10% (w/w) at a pressure level of
200 atm.

Figure 2 assembles more data on the solubility of naphthalene
in carbon dioxide from the previously mentioned works which was
assembled by Modell (26) and the data are plotted in a manner which
will aid in the explanation of how a supercritical fluid extraction
process operates. The curves in the figure give isobars of
solubility as a function of temperature, and the dotted curve shows
the solubility of naphthalene in liquid carbon dioxide up to the
critical temperature of 31°C and in saturated vapor carbon dioxide.
As one other example of a system for which a substantial amount of
data has been obtained, Figure 3 shows the solubility behavior of
triglycerides in supercritical carbon dioxide (27,28). The
absolute values, the pressure, and temperature levels are
different, but the characteristic "fan" of curves is similar in
shape to those in Figure 2, which partially lends credence to the
statement that the behavior of many supercritical
solvent–incompletely miscible solute binary systems is similar. As
an additional example, Figure 4 shows the solubility behavior of a
very different solute–solvent system, silica–water (29). Comparison
of the solubility levels with temperature and pressure levels again
points out that the absolute values for all systems are not the
same; however, the general shapes of the curves of all systems,
i.e., the "fans" referred to earlier, are the same. Many other
solubility diagrams can be constructed for other solute–solvent
pairs; for example, phenanthrene (30), benzoic acid (31),
anthracene (32), phenol and chlorinated phenols (33), biphenyl
(21), represent just a small list of solutes that have been studied
for their solubility and phase behavior in methane, ethylene,
ethane, and carbon dioxide. They all exhibit the similarity of the
data shown in Figures 2, 3, and 4.

Figure 2, showing the solubility behavior of naphthalene in
supercritical carbon dioxide, and Figures 3 and 4, which point out
that the behavior is quite general, will be used to explain how a
"generic" supercritical fluid extraction process operates.

Operation of a Supercritical Fluid Extraction Process

Because of the dissolving characteristics shown in Figures 1–4, it
is possible to design industrial processes to extract, purify, and
fractionate materials based on changes in pressure of a
supercritical fluid solvent, at high pressure effecting an

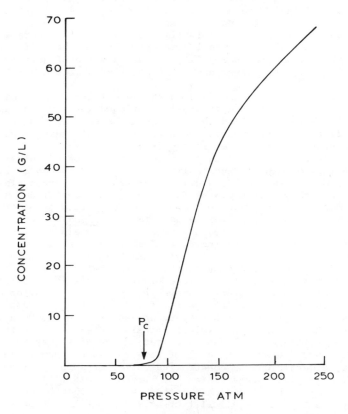

Figure 1.   Solubility of Naphthalene in Carbon Dioxide at 45°C.

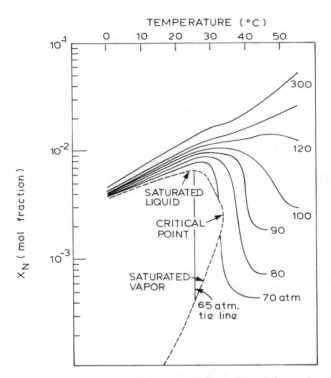

Figure 2. Solubility of Naphthalene in Carbon Dioxide.

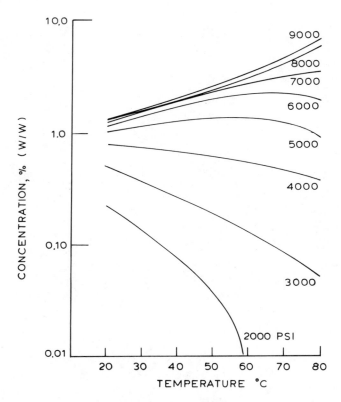

Figure 3.   Solubility of Triglycerides in Carbon Dioxide.

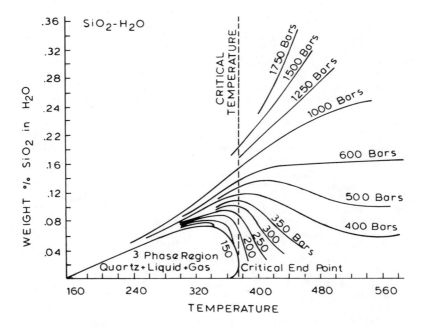

Figure 4.   Solubility of Silica in Water.

extraction and at a lower pressure effecting a separation of the
dissolved material from the solvent which can then be recycled to
the extractor for further extraction. The process can be either
batch or continuous depending upon the nature of the feed and the
nature of the extraction, i.e., whether it be a purification or
fractionation, or an extraction from a reaction mass.

A schematic diagram of a process that uses a supercritical
fluid as a pressure-dependent solvent to extract an organic
substance is given in Figure 5a. Four basic elements of the
process are shown, viz., an extraction vessel, a pressure reduction
valve, a separator for collecting the material dissolved in the
extractor, and a compressor for recompressing and recycling fluid.
Ancillary pumps, valving, facilities for filling the vessel and for
fluid make-up, heat exchangers, and similar equipment are omitted
from the figure for clarity and ease of presentation. Figure 5b
shows the extensive data on solubility of naphthalene in carbon
dioxide as a function of temperature and pressure given previously
in Figure 2.

Some process operating parameters are indicated on two
solubility isobars in Figure 5b; E_1 represents conditions in the
extractor, e.g., 300 atm, 55°C, and S_1 the conditions which exist
in the separator, 90 atm, 43°C. The extractor vessel is assumed to
be filled with naphthalene in admixture with another material,
which for ease of discussion is assumed to be insoluble in carbon
dioxide. Carbon dioxide at condition E_1 is passed through the
extraction vessel wherein it extracts the naphthalene from the
insoluble material. The carbon dioxide-naphthalene solution
leaving the extractor is expanded to 90 atm through the pressure
reduction valve and as indicated by the directed path in Figure 5b.
Because the equilibrium solubility has been reduced from about 5%
to 0.2% during the pressure reduction step, naphthalene
precipitates from the solution. The naphthalene is collected in
the separator, and the carbon dioxide leaving the separator is
recompressed and returned to the extractor. Recycling continues
until all the naphthalene is extracted. The directed line segment
E_1-S_1 in Figure 5b and its reverse represent approximately the
cyclic process on the solubility diagram.

As an alternative to extraction and separation using pressure
reduction (i.e., the path E_1-S_1), the process can operate at
constant pressure with temperature changes in a supercritical fluid
used to effect the extraction and separation steps. For example,
and again starting at E_1, the stream leaving the extractor could be
passed through a heat exchanger (instead of a pressure reduction
valve) and cooled to, for example, 20°C as indicated by the
directed portion on the 300 atm isobar. Cooling the stream would
result in the precipitation and collection of naphthalene in the
separator at conditions S_2. The carbon dioxide leaving the
separator could then be heated back to 55°C before being recycled
to the extractor. This mode of operation, instead of requiring a
high ΔP compressor would employ a low ΔP and less expensive
blower to supply frictional loss which is small.

Although not specifically accented on the solubility diagram,
there are regions where the solubility decreases with increasing
temperature, a behavior not usually exhibited by liquid solvents,

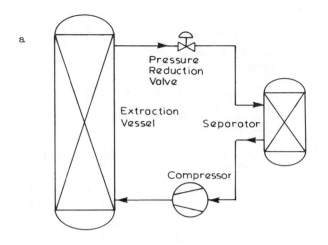

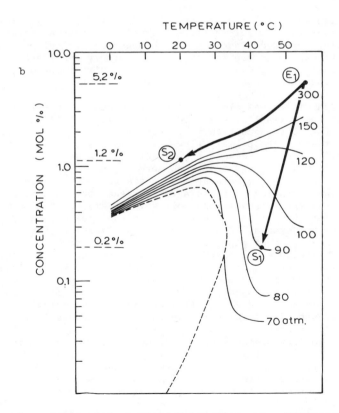

Figure 5.   Supercritical Fluid Extraction – a.   Process Diagram;
b.   Operating Paths.

and this behavior can be used to advantage in carrying out a supercritical fluid extraction and separation. For example, one could carry out the extraction of naphthalene at 30°C and 80 atm and cause a 10-fold change in dissolving power by <u>heating</u> the solution only 5 or 10°C.

At the end of the extraction cycle, essentially all the naphthalene is in the separator and the insoluble material is left in the extractor. The process can be considered as a naphthalene extraction or as an "insolubles" purification depending upon which is the desired product.

Applications of Supercritical Fluids to the Extraction and Characterization of Flavors

In addition to the ability of carbon dioxide to act as a lipophilic solvent, its dissolving power can be tailored to carry out fractionations and relative separations of components in a complex mixture, and this unique property is advantageously applied to the characterization of flavors. Several representative examples that point out this ability are presented.

Ginger. Figure 6 is a chromatogram obtained by capillary column gas chromatography (GC) of an extract of ginger that was obtained by extracting shredded ginger root with methylene chloride in a Soxhlet extractor. As can be seen from a cursory examination of the chromatogram, ginger extract is a complex mixture of many components (no attempt was made to identify any of them for this presentation).

Shredded ginger was also extracted with supercritical carbon dioxide and the extract(s) analyzed by GC. Figure 7 is a schematic diagram of the continuous flow laboratory apparatus used for the extraction tests. The primary elements are a gas supply, compressor, extraction vessel, flow control and pressure reduction valve, gas flow meter, and total gas meter, plus such ancillary equipment as heaters, temperature and pressure control, etc., which are not shown in the schematic diagram for ease of discussion. The conduct of the extraction was as follows: An amount of ginger was introduced into the pressure vessel, the vessel sealed, and connected to the system. (The pressure vessels are typically 2" nominal, SCh 160,316 st pipe, 1 to 4 ft long, 1/2 to 2 liters volume.) The temperature of the system was brought to the desired temperature, and carbon dioxide from a manifold and available at about 1200 psi was fed to the suction side of a diaphram compressor. The compressed carbon dioxide at the desired pressure was passed through the ginger-filled extractor, and soluble materials present in the ginger were dissolved. The solution leaving the extractor was passed through the flow control valve and expanded to ambient pressure which caused the dissolved materials to nucleate and precipitate in the collector. The 1 atm gas passed subsequently through the flow meter and dry test meter. Gas volume data coupled with gravimetric determinations allow concentrations, yields, rates of extraction, and similar information to be calculated.

Figure 8 is a chromatogram of the extract obtained by

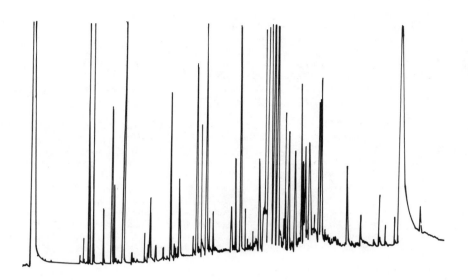

Figure 6.   Gas Chromatogram - Methylene Chloride Extract of Ginger.

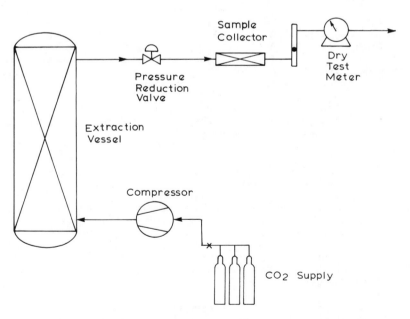

Figure 7.   Laboratory Apparatus for Supercritical Fluid Extraction.

extracting ginger with supercritical carbon dioxide at conditions of 5000 psi and 50°C. For the qualitative purpose of this discussion, the chromatograms shown in Figure 6 and Figure 8 are essentially the same, which shows that 5000 psi carbon dioxide is comparable in its extraction characteristics to methylene chloride. However, the physical appearance of the two extracts was slightly different, viz., the carbon dioxide extract was a yellow-green, thin paste, whereas the methylene chloride extract was darker green.

In order to illustrate the pressure dependent dissolving power of a supercritical fluid, another charge of shredded ginger was extracted with carbon dioxide at two other pressure levels. The charge was first extracted at a low pressure, 1500 psi and 50°C. The extract that was collected was a pale yellow liquid. The pressure was then raised to a medium pressure level of 3000 psi; and another volume of gas passed through the same charge of shredded ginger. The extract that was collected was yellow-green in color, but a thicker paste than that obtained in the previous test carried out at 5000 psi.

Figure 9 is a chromatogram of the 1500 psi extract and Figure 10 the chromatogram of the 3000 psi extract; Figure 9 is clearly different from the chromatogram of the 5000 psi extract shown in Figure 8 in the number and ratios of peaks. A cursory comparison of Figures 8, 9, and 10 reveals that carbon dioxide at 1500 psi does not extract many of the longer retention components (i.e., those peaks to the right of the chromatograms). A closer examination of the chromatograms shows that 1500 psi carbon dioxide does extract some of the long retention time compounds, while it does not extract all the low retention time components.

The relation between the retention time on a chromatographic column and the solubility in carbon dioxide was not determined in this study, and there probably is no monotonic relation. Retention time may reflect more than say a vapor pressure phenomenon and can be a result of polarity considerations. In a homologous series, for example, the retention time on a column is related to solely vapor pressure, and analogously the solubility in a supercritical fluid would exhibit a monotonic relationship, viz., the higher the vapor pressure (and therefore the lower the retention time), the higher the solubility. However, across families of compounds, for example, organic acids, ketones, esters and aldehydes, there is probably not a general correlation between retention time on a specific column packing and vapor pressure, and similarly there would not be a correlation between retention time and the solubility in a supercritical fluid. Figure 9 does show, nevertheless, that some long retention time components are extracted while other low retention time materials are not; it is offered that the mixture characterized by the chromatogram in Figure 9 would be difficult to achieve by distillation or by liquid solvent extraction, and one of the advantages of supercritical fluid extraction lies in the ability to isolate or fractionate certain groups of compounds for subsequent evaluation or characterization.

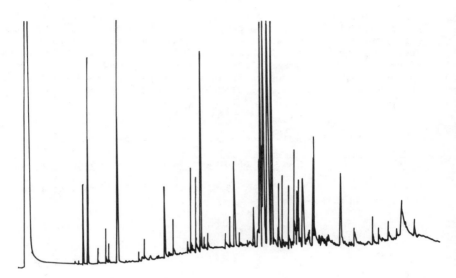

Figure 8.   Gas Chromatogram – Carbon Dioxide Extract (5000 psi) of Ginger.

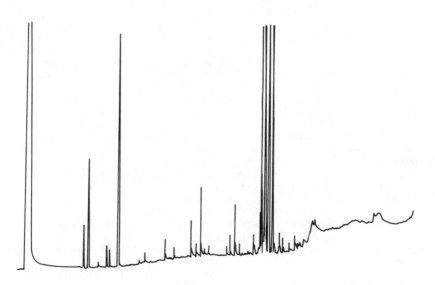

Figure 9.   Gas Chromatogram – Carbon Dioxide Extract (1500 psi) of Ginger.

Figure 10 shows the chromatogram of the 3000 psi extract of ginger described earlier; a comparison of Figures 9 and 10 and the high pressure extract whose chromatogram was given in Figure 8 shows that the extract taken sequentially after first stripping the ginger at 1500 psi was depleted of some of the low retention time components that were concentrated at 1500 psi as shown in Figure 9. The gravimetric analysis of all the extracts is given in Table I.

Table I. Gravimetric Analysis of Ginger Extracts

Test	Amount of Ginger Charged	Amount of Extract
Soxhlet, methylene chloride	5.07 g	Not determined
5000 psi, carbon dioxide	30.24 g	2.74 g
1500 psi, carbon dioxide	32.44 g	0.35 g
3000 psi, carbon dioxide	Same charge as above	1.74 g

Pimento Berries ("All Spice"). A similar set of extraction tests was carried out with crushed pimento berries, and qualitatively similar results were obtained. Figure 11 is a chromatogram of the methylene chloride extract of pimento berries, and no commentary on the complexity of the chromatogram is needed. Figure 12 is the high pressure carbon dioxide extract of pimento berries and the two chromatographs are similar in appearance. Extraction conditions with the carbon dioxide were again 5000 psi, 50°C.

A second charge of crushed pimento berries was extracted sequentially at the same two pressure levels tested for the ginger, 1500 psi and 3000 psi, and the chromatograms of the two extracts are shown in Figures 13 and 14, respectively. The low pressure extract was almost water-white, and the 3000 and 5000 psi extracts were yellow to yellow-green pastes. A comparison of Figures 12, 13, and 14 allows the same qualitative conclusions to be drawn as they were for the ginger extractions, viz., low pressure carbon dioxide is different in its extractive capabilities than is high pressure carbon dioxide, and based upon the appearance of the chromatogram 5000 psi carbon dioxide extracts the same components that methylene chloride does. Additionally, the sequential 3000 psi extract is depleted in some of the soluble components that were concentrated in the 1500 psi extract.

For completeness, Table II gives the gravimetric analysis of the extracts of pimento berries.

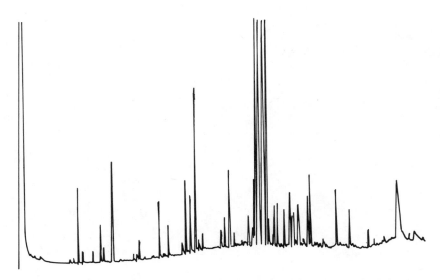

Figure 10.   Gas Chromatogram – Carbon Dioxide Extract (3000 psi) of Ginger.

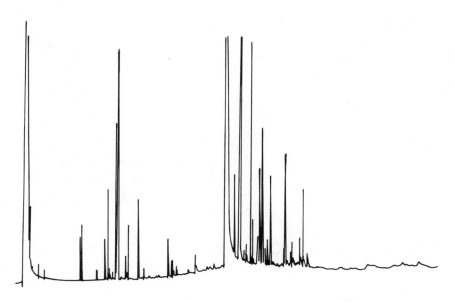

Figure 11.   Gas Chromatogram – Methylene Chloride Extract of Pimento Berries.

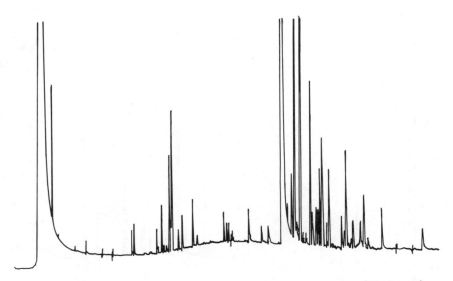

Figure 12. Gas Chromatogram – Carbon Dioxide Extract (5000 psi) of Pimento Berries.

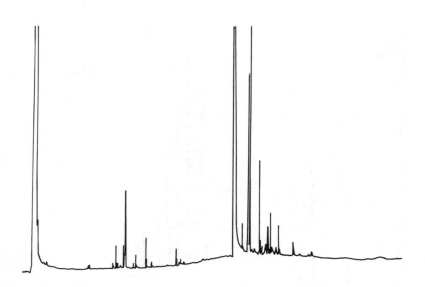

Figure 13. Gas Chromatogram – Carbon Dioxide Extract (1500 psi) of Pimento Berries.

Table II. Gravimetric Analysis of Pimento Berry Extract

Test	Amount of Pimento Berries	Amount of Extract
Soxhlet	4.34 g	Not determined
5000 psi	25.97 g	2.41 g
1500 psi	27.41 g	0.32 g
3000 psi	Same charge as above	1.64 g

Extraction/Concentration of Flavors from Solutions. Carbon dioxide
is also an ideal solvent for concentrating essence components, and
one example of this capability is presented in this subsection.
Figure 15 is a chromatogram of a synthetic apple essence consisting
of eight C_6-C_8 esters, alcohols, and aldehydes dissolved in water.
Extraction was carried out in the same laboratory apparatus shown
in Figure 7 with the differences that the vessel was fitted with a
check valve to prevent escape of the solution and with a flow
distributor sieve tray to enhance contact between the solution and
the carbon dioxide flowing through the charge of liquid.

An amount of solution (302 g in this test) was contacted with
800 g of carbon dioxide in the batch-continuous mode described
earlier, and the extract was collected downstream of the expansion
valve in a trap maintained at dry ice temperature to capture the
volatile components. A chromatogram of the extract is given in
Figure 16. A quick comparison of Figures 15 and 16 does not
provide qualitative differences to be discerned because the peaks
rise off-scale in both cases. The feed solution was composed of
eight components, each present at about 0.02% (for a "total"
concentration of 0.19%). Each of the two peaks indicated by arrows
in both the starting essence and the extract actually consists of
two components which were resolved in the GC analysis (note the
double, almost-superposed time print above each peak) but because
of the scale selected they are not visually identifiable as
separate peaks on the chromatogram. The integrator output from the
GC analysis of the extract, however, showed that the total
concentration in the extract was 12.4% (an average of about 1.5%
per component). Further inspection of Figures 15 and 16 shows that
some trace components (market T in both figures) are concentrated
in the extract. These trace components were present in the
compounds used to prepare the synthetic apple essence and are
related aldehydes, acids, etc.

The chromatogram of the raffinate, the depleted solution from
the extraction, is given in Figure 17, and comparison provides
interesting qualitative results. First, all the component
concentrations have been reduced substantially (which is not
surprising because high carbon number lipophilic materials are
readily extracted by carbon dioxide). However, the peak heights
(and, more quantitatively, the integrator output not reproduced
here) show that not all the components are extracted equally.
Additionally, the two peaks marked by arrows can now be seen to
contain two components each. As was found for the previously
discussed ginger and pimento berries extractions, not all the apple
essence components are extracted equally. Shultz et al (34) have

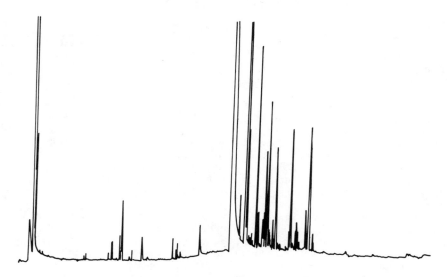

Figure 14. Gas Chromatogram - Carbon Dioxide Extract (3000 psi) of Pimento Berries.

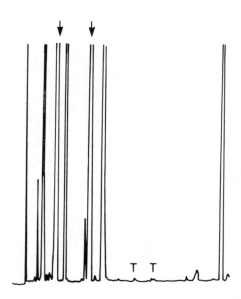

Figure 15. Gas Chromatogram - Synthetic Apple Essence.

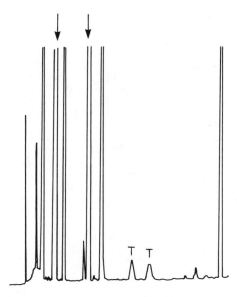

Figure 16. Gas Chromatogram – Carbon Dioxide Extract of Apple Essence.

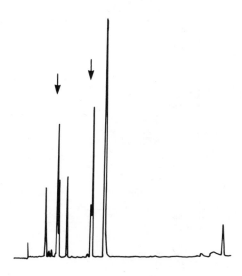

Figure 17. Gas Chromatogram – Depleted Apple Essence.

shown, for example, that esters exhibit higher distribution coefficients than aldehydes and carboxylic acids in organic–water–carbon dioxide systems.

In closing, this paper was not intended to represent an exhaustive process development effort in flavors extraction from natural materials nor a development of the quantitative analytical capabilities of supercritical carbon dioxide. However, even though the examples and the conditions of extraction were somewhat arbitrary, they point out some of the interesting features of the pressure dependent dissolving power properties of supercritical fluids. They can be further refined by virtue of more narrow ranges and ratios of pressure and temperature to accomplish still more narrow separations.

On the other hand, the historical perspective and operation of supercritical extraction was intended to supply the background to serve as an aid in appreciating the motivation for the current activities, not just in flavors extraction and characterization but for the process development being carried out in the chemical, food, pharmaceutical, and polymer industries.

Literature Cited

1. Krukonis, V. J., Branfman, A. R., and Broome, M. G. 87th Nat. Mtg.; AIChE: Boston, August 1979.
2. Hannay, J. B., Hogarth, J. Proc. Roy. Soc. (London) 1879, 29, 324–6.
3. Ramsay, W. Proc. Roy. Soc. (London) 1880, 30, 323–9.
4. Hannay, J. B. and Hogarth, J. Proc. Roy. Soc. (London) 1880, 30, 178–88.
5. Hannay, J. B. Proc. Roy. Soc. (London) 1880, 30, 484–9.
6. Booth, H. S. and Bidwell, R. M. Chem. Rev. 1949, 44, 477–513.
7. Irani, C. A., and Funk, E. W. In CRC Handbook: Recent Developments in Separation Science; CRC Press, Boca Raton, Florida, Vol. III, Part A, 1977; pp. 171–9.
8. Williams, D. F. Chem. Eng. Sci. 1981, 36, 1769–88.
9. Paulaitis, M. E., Krukonis, V. J., Kurnik, R. T., and Reid, R. C. Rev. in Chem. Eng. 1983, 1, 179–250.
10. Modell, M., deFilippi, R. P., and Krukonis, V. J. 87th Nat. Mtg.; AIChE: Boston, August 1979.
11. Moses, J. M., Goklen, K. E., and deFilippi, R. P. 1982 Ann. Mtg.; AIChE: Los Angeles, November.
12. Caragay, A. B. Perfume and Flavorist 1981, 6, 43–55.
13. Schultz, E. G., and Randal, J. N. Food Tech 1970, 24, 94–8.
14. Diepen, G. A. M. and Scheffer, F. E. C. J. Am. Chem. Soc. 1948, 70, 4081–5.
15. Diepen, G. A. M. and Scheffer, F. E. C. J. Am. Chem. Soc. 1948, 70, 4085–9.
16. Diepen, G. A. M. and Scheffer, F. E. C. J. Phys. Chem. 1953, 57, 575–8.

17. Tsekhanskaya, Yu. V., Iomtev, M. B. and Mushkina, E. V.
     Russ. J. Phys. Chem. 1964, 38, 1173-6.
18. Tsekhanskaya, Yu. V., Iomtev, M. B. and Mushkina, E. V.
     Russ. J. Phys. Chem. 1962, 36, 1177-81.
19. King, A. D., and Robertson, W. W. J. Chem. Phys. 1962, 37,
     1453-5.
20. Najour, G. C. and King, A. D., Jr., J. Chem. Phys. 1966,
     45, 1915-21.
21. McHugh, M. A., and Paulaitis, M. E. J. Chem. Eng. Data
     1980, 25, 326-9.
22. Kurnik, R. T. and Reid, R. C. Fluid Phase Equilibria 1982,
     8, 93-105.
23. Schmitt, W. J. and Reid, R. C. 1984 Ann. Mtg.; AIChE: San
     Francisco, November.
24. Krukonis, V. J., McHugh, M. A., Seckner, A. J. J. Phys.
     Chem. 1984, 88, 2687-9.
25. McHugh, M. A., Seckner, A. J. and Krukonis, V. J. 1984 Ann.
     Mtg.; AIChE: San Francisco, November.
26. Modell, M., deFilippi, R. P., Krukonis, V. J. and Robey, R. J.
     87th AIChE Mtg., Boston, August 1979.
27. Friedrich, J. P., List, G. R. and Spencer, G. F. 75th Am.
     Oil Chem. Soc. Mtg., Dallas, May 1984.
28. Krukonis, V. J. 75th Am. Oil Chem. Soc. Mtg., Dallas, May
     1984.
29. Kennedy, G. C. Econ. Geol. 1950, 45, 629-36.
30. Eisenbeiss, J. A Basic Study of the Solubility of Solids in
     Gases at High Pressures, Final Report, Contract No.
     DA18-108-AMC-244(A), Southwest Research Inst., San Antonio,
     August 6, 1984.
31. Kurnik, R. T., Holla, S. J. and Reid, R. C. J. Chem. Eng.
     Data. 1981, 26, 47-51.
32. Najour, G. C. and King, A. D., Jr. J. Chem. Phys. 1970, 52,
     5206-11.
33. VanLeer, R. A., and Paulaitis, M. E. J. Chem. Eng. Data
     1980, 25, 257-9.
34. Shultz, W. G. and Randall, J. N. 1970, 24, 94-8.

RECEIVED September 9, 1985

Author Index

Subject Index

Production by Meg Marshall
Indexing by Karen McCeney
Jacket design by Pamela Lewis

Elements typeset by Hot Type Ltd., Washington, D.C.
Printed and bound by Maple Press Co., York, Pa.